Hydrogeology and Water Quality of the Floridan Aquifer System and Effect of Lower Floridan Aquifer Pumping on the Upper Floridan Aquifer, Pooler, Chatham County, Georgia, 2011–2012

By Gerard J. Gonthier

Prepared in cooperation with the City of Pooler, Georgia

Scientific Investigations Report 2012–5249

U.S. Department of the Interior
U.S. Geological Survey

U.S. Department of the Interior
KEN SALAZAR, Secretary

U.S. Geological Survey
Marcia K. McNutt, Director

U.S. Geological Survey, Reston, Virginia: 2012

For more information on the USGS—the Federal source for science about the Earth, its natural and living resources, natural hazards, and the environment, visit *http://www.usgs.gov* or call 1-888-ASK-USGS

For an overview of USGS information products, including maps, imagery, and publications, visit *http://www.usgs.gov/pubprod*

To order this and other USGS information products, visit *http://store.usgs.gov*

Suggested citation:
Gonthier, G.J., 2012, Hydrogeology and water quality of the Floridan aquifer system and effect of Lower Floridan aquifer pumping on the Upper Floridan aquifer, Pooler, Chatham County, Georgia, 2011–2012: U.S. Geological Survey Scientific Investigations Report 2012–5249, 62 p., available at *http://pubs.usgs.gov/sir/2012/5249/*.

Contents

Figures

Appendix Figures

1–5 to 2–4. Graphs showing—

Tables

Appendix Tables

Conversion Factors and Datum

Inch/Pound to SI

Multiply	By	To obtain
Length		
inch	25.4	millimeter (mm)
mile (mi)	1.609	kilometer (km)
Volume		
cubic foot (ft^3)	0.02832	cubic meter (m^3)
acre-foot (acre-ft)	0.325851	million gallons (Mgal)
Flow rate		
inch per year (in/yr)	25.4	millimeter per year (mm/yr)
foot per day (ft/d)	0.3048	meter per day (m/d)
foot squared per day (ft^2/day)	0.092903	meter squared per day (m^2/d)
gallon per minute (gal/min)	0.06309	liter per second (L/s)
million gallons per day (Mgal/d)	0.0003785	million cubic meters per day (Mm^3/d)

Transmissivity: The standard unit for transmissivity is cubic foot per day per square foot times foot of aquifer thickness [(ft^3/d)/ft^2]ft. In this report, the mathematically reduced form, foot squared per day (ft^2/d), is used for convenience.

Temperature in degrees Fahrenheit (°F) may be converted to degrees Celsius (°C) as follows:

$$°C = (°F - 32) / 1.8$$

Specific conductance is given in microsiemens per centimeter at 25 degrees Celsius (µS/cm at 25 °C). In this report, microsiemens per centimeter (µS/cm) is used for convenience.

Vertical coordinate information is referenced to North American Vertical Datum of 1988 (NAVD 88).

Altitude, as used in this report, refers to distance above the vertical datum.

Abbreviations

ASTM	American Society for Testing and Materials
EM	electromagnetic
ft^2/d	foot squared per day
GaEPD	Georgia Environmental Protection Division
gal/d	gallon per day
gal/min	gallon per minute
K_h	horizontal hydraulic conductivity
K_v	vertical hydraulic conductivity
LFA	Lower Floridan aquifer
LFCU	Lower Floridan confining unit
MAX	maximum drawdown
NOAA	National Oceanic and Atmospheric Administration
NTU	nephelometric turbidity unit
PCU	platinum-cobalt unit
RMS	root mean square
SI	sample interval
UFA	Upper Floridan aquifer
USGS	U.S. Geological Survey
μg	microgram
μm	micrometer
μg/L	microgram per liter
mg/d	milligram per day
mg/L	milligram per liter

Acknowledgments

The following individuals provided abundant cooperation with the U.S. Geological Survey (USGS): Matt Saxon and Mark Williams of the Pooler Public Works Department provided helpful support throughout the study. Harry Jue and Holmes Bell of Hussey, Gay, Bell & DeYoung coordinated drilling and USGS testing activities at the test site. Rowe Drilling Company, Inc., drilled two boreholes and completed them as wells, provided and operated the packer assembly, and installed the test pump and provided assistance monitoring two aquifer performance tests. TestAmerica Laboratories, Inc., performed water chemistry analyses on collected water-quality samples and S&ME, Inc., performed hydraulic analyses on cores.

The author extends his appreciation for the dedicated work provided by O. (Gary) Holloway and Michael D. Hamrick, USGS. Gary Holloway fabricated the slug used for slug tests, designed and constructed the aquifer-test plumbing to remove discharge water from the site, and measured pump-discharge rate. Gary Holloway and Michael Hamrick provided the teamwork for monitoring drill-site activities and collecting hydrogeologic and water-quality data, which included drill cuttings, core samples, geophysical logging, flowmeter surveys, water-quality samples, and continuous and manual water-level measurements before, during, and after aquifer tests.

Hydrogeology and Water Quality of the Floridan Aquifer System and Effect of Lower Floridan Aquifer Pumping on the Upper Floridan Aquifer, Pooler, Chatham County, Georgia, 2011–2012

By Gerard J. Gonthier

Abstract

Two test wells were completed in Pooler, Georgia, in 2011 to investigate the potential of using the Lower Floridan aquifer as a source of water for municipal use. One well was completed in the Lower Floridan aquifer at a depth of 1,120 feet (ft) below land surface; the other well was completed in the Upper Floridan aquifer at a depth of 486 ft below land surface. At the Pooler test site, the U.S. Geological Survey performed flowmeter surveys, packer-isolated slug tests within the Lower Floridan confining unit, slug tests of the entire Floridan aquifer system, and aquifer tests of the Upper and Lower Floridan aquifers.

Drill cuttings, geophysical logs, and borehole flowmeter surveys indicate that the Upper Floridan aquifer extends 333–515 ft below land surface, the Lower Floridan confining unit extends 515–702 ft below land surface, and the Lower Floridan aquifer extends 702–1,040 ft below land surface.

Flowmeter surveys indicate that the Upper Floridan aquifer contains two water-bearing zones at depth intervals of 339–350 and 375–515 ft; the Lower Floridan confining unit contains one zone at a depth interval of 550–620 ft; and the Lower Floridan aquifer contains five zones at depth intervals of 702–745, 745–925, 925–984, 984–1,015, and 1,015–1,040 ft. Flowmeter testing of the test borehole open to the entire Floridan aquifer system indicated that the Upper Floridan aquifer contributed 92.4 percent of the total flow rate of 708 gallons per minute; the Lower Floridan confining unit contributed 3.0 percent; and the Lower Floridan aquifer contributed 4.6 percent.

Horizontal hydraulic conductivity of the Lower Floridan confining unit derived from slug tests within three packer-isolated intervals ranged from 0.5 to 10 feet per day (ft/d). Aquifer-test analyses yielded values of transmissivity for the Upper Floridan aquifer, Lower Floridan confining unit, and the Lower Floridan aquifer of 46,000, 700, and 4,000 feet squared per day (ft^2/d), respectively. Horizontal hydraulic conductivity of 4 ft/d for the Lower Floridan confining unit, derived from aquifer-test analyses, is near the midrange for values derived from packer-isolated slug tests. The transmissivity of the entire Floridan aquifer system derived from aquifer-test analyses totals about 51,000 ft^2/d, similar to the value of 58,000 ft^2/d derived from open slug tests on the entire Floridan aquifer system.

Water-level data for each aquifer test were filtered for external influences such as barometric pressure, earth-tide effects, and long-term trends to enable detection of small (less than 1 foot) water-level responses to aquifer-test pumping. During the 72-hour aquifer test of pumping the Lower Floridan aquifer, a drawdown response of 51.7 ft was observed in the Lower Floridan pumped well and a drawdown response of 0.9 foot was observed in the Upper Floridan observation well located 85 ft from the pumped well.

Introduction

The City of Pooler is located in western Chatham County, Georgia (Ga.), near the City of Savannah (fig. 1). Public water supply within the city is predominantly derived from groundwater withdrawn from two wells (36Q283 and 36Q348), completed in the Upper Floridan aquifer (UFA). Concern over saltwater intrusion at Hilton Head Island, South Carolina, has resulted in increased restrictions on groundwater withdrawal from the UFA by the Georgia Environmental Protection Division (GaEPD) in the Chatham County area. To meet the growing water demand in the 24-county coastal Georgia area, GaEPD has encouraged usage of alternative sources of water to the UFA, including wells completed in the Lower Floridan aquifer (LFA). The City of Pooler seeks to use the LFA for municipal water use.

Pumping from the LFA may locally increase the vertical hydraulic head gradient between the UFA and LFA, induce leakage (groundwater flow) from the UFA to the LFA, and lower water levels in the UFA. As a result, the GaEPD requires an assessment of these effects as a permitting requirement. In January 2003, GaEPD released an interim strategy for permitting LFA groundwater withdrawals in the 24-county coastal Georgia area (Nolton Johnston, Georgia Environmental Protection Division, written commun., January 28, 2003).

To assess the water-supply potential of the LFA at Pooler, Ga., the U.S. Geological Survey (USGS) in cooperation with the City of Pooler, performed an investigation during 2011–2012 to determine the hydrogeology and water quality of the Floridan aquifer system and the potential effect that pumping from the LFA would have on the UFA. The study included construction of a test well in the UFA and a test well in the LFA, detailed site investigations, and hydraulic characterization of the Floridan aquifer system.

Purpose and Scope

This report documents results of field investigations completed at Pooler, Ga., during 2011–2012 to determine the hydrogeology and water quality of the Floridan aquifer system and to provide data needed to assess the effect of LFA pumping on the UFA, specifically to:

- Determine hydraulic and water-quality characteristics of the UFA, LFA, and the intervening Lower Floridan confining unit (LFCU), and

- Identify how pumping the LFA affects water levels in the UFA.

Field investigations included:

- Boring a 1,131-foot (ft)-deep test hole and constructing a 1,120-ft-deep test well completed in the LFA;

- Collecting drill cuttings and borehole geophysical logs at the test well;

- Sampling core within the LFCU for analysis of vertical hydraulic conductivity (K_v) and porosity;

- Performing flowmeter surveys throughout the Floridan aquifer system in the open test hole and in the completed LFA well;

- Performing slug tests in packer-isolated intervals in the open test hole within the LFCU, and performing slug tests on the test hole open to the entire Floridan aquifer system;

- Collecting depth-integrated water samples to assess water quality of various water-bearing zones;

- Boring and constructing a 486-ft-deep observation well completed in the UFA; and

- Performing a 24-hour aquifer test at the test well open to the UFA, and a 72-hour aquifer test at the test well open to the LFA.

Data and subsurface samples collected during field investigations facilitated a hydrogeologic description of the subsurface at the test-well site by (1) determining transmissivity of the LFA, LFCU, and UFA; (2) storage coefficient of the LFCU and UFA; (3) specific capacity of the LFA well; and (4) horizontal hydraulic conductivity (K_h) and K_v of the LFCU. In conjunction with digital simulation and further analyses, this information can aid in determining the amount of pumping reduction in the UFA required by GaEPD to offset drawdown and leakage resulting from pumping a new LFA well located at Pooler, Ga. Results of this investigation add to the body of knowledge needed to characterize the Floridan aquifer system on a regional basis.

Previous Studies

Recent hydrologic investigations by the USGS, in cooperation with the U.S. Army, characterized the hydrogeology and groundwater flow in the Floridan aquifer system at Hunter Army Airfield in Chatham County, Ga., about 7 miles (mi) southeast of the Pooler test site (Clarke and others, 2010; Williams, 2010), and at Fort Stewart in Liberty County, Ga., about 23 mi southwest of the Pooler test site (Clarke and others, 2011; Gonthier, 2011). These investigations included data collection and groundwater-model simulations to determine the hydrogeology and water quality of the Floridan aquifer system and to provide information needed to assess the effect of LFA pumping on the UFA. The method of study of the current investigation at the Pooler test site is identical to the two earlier investigations at Hunter Army Airfield and Fort Stewart.

A revised hydrogeologic framework for the Floridan aquifer system was developed by Williams and Gill (2010) for eight northern coastal counties in Georgia and five coastal counties in South Carolina, including the area surrounding Pooler, Ga. Borehole-geophysical and flowmeter-survey logs collected during previous investigations were used by Williams

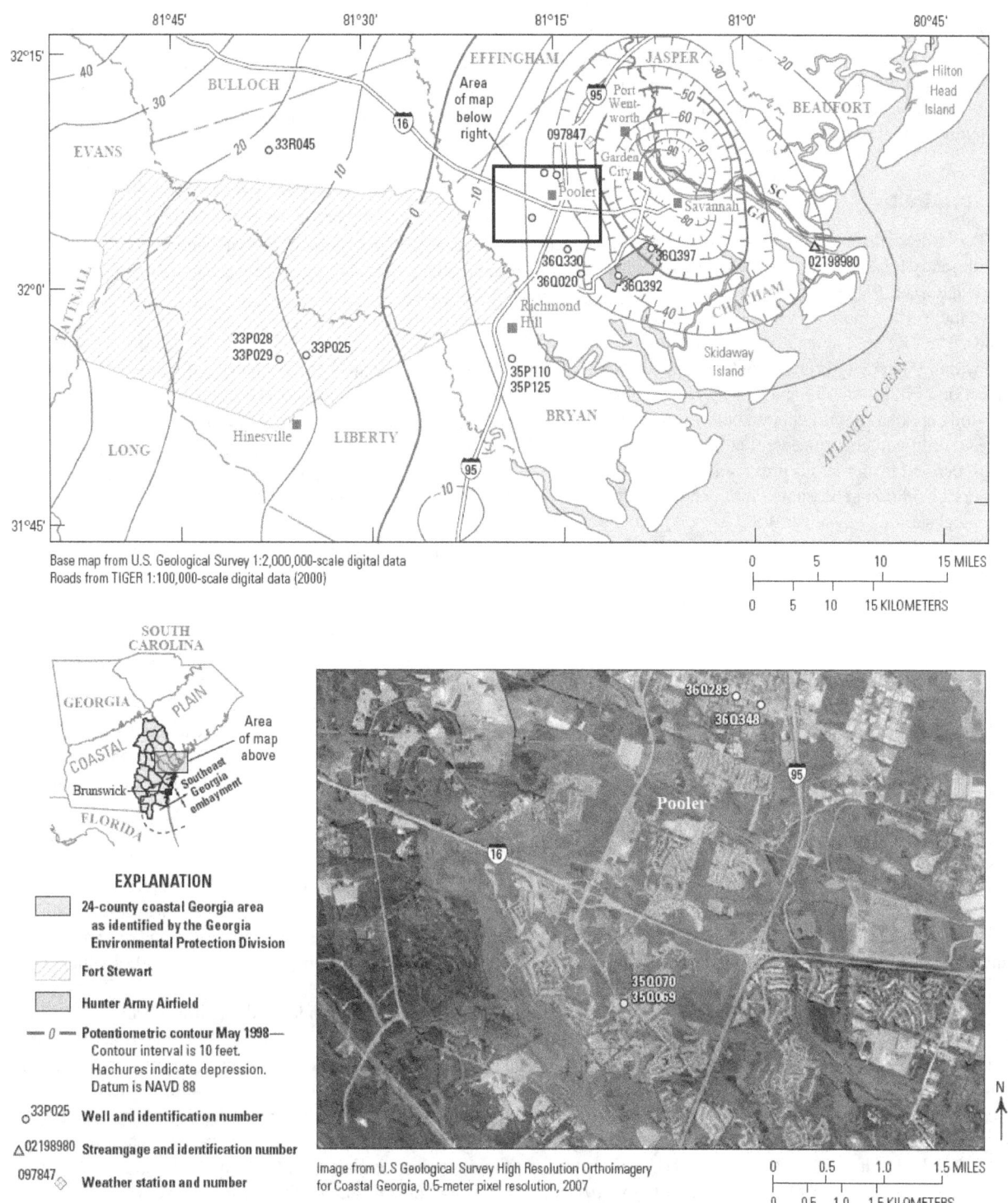

Figure 1.   Location of test site (wells 35Q069 and 35Q070), Pooler, Georgia.

and Gill (2010) to shift the position of hydrogeologic vertical boundaries of the Upper and Lower Floridan aquifers and of individual permeable zones contained within these aquifers. The investigation included a deep test well (36Q330) drilled at Berwick Plantation in Chatham County, Ga., located 3.73 mi southeast of the Pooler test site, which provided data that helped correlate hydrogeologic-unit depths at the Pooler test site.

Site Description

The Pooler test site is characterized by flat topography capped with sandy topsoil typical of the Georgia coastal area (Clark and Zisa, 1976). Land-surface altitude is about 19 ft above the North American Vertical Datum of 1988 (NAVD 88). The LFA well at the Pooler test site (well 35Q069) is located within Pooler city limits, about 1.7 mi southwest of the intersection of Interstates 16 and 95 (fig. 1).

The study area has a mild climate with warm, humid summers and mild winters. Long-term climatic patterns in the area are derived from records provided by the National Weather Service Station at Savannah International Airport (climatological station "Savannah WSO Airport, Georgia [097847]," accessed at *http://www.sercc.com/cgi-bin/sercc/cliMAIN.pl?ga7847*, on September 27, 2011). During 1971–2000, precipitation at station 097847 averaged about 49 inches per year (in/yr).

Water Use

Groundwater use in Chatham County totaled 64.97 million gallons per day (Mgal/d) during 2005 (Fanning and Trent, 2009). About 50 percent (33.52 Mgal/d) was withdrawn for public supply. Groundwater is supplied by about 60 wells in Chatham County (Fanning and Trent, 2009). Although most of these wells are located in Savannah, several are located to the northwest, serving Pooler, Garden City, and Port Wentworth (Fanning and Trent, 2009). Payne and others (2005) estimated that during 1980–2000, close to 95 percent of groundwater withdrawn from the county was derived from the UFA, with the remaining 5 percent obtained from the LFA. Groundwater withdrawal from the Floridan aquifer system in Chatham County increased from 79.75 Mgal/d in 1980 to 85.54 Mgal/d in 1990, and decreased to 68.15 Mgal/d in 2000 (Payne and others, 2005).

Water demand for the City of Pooler has increased with increasing population. Pooler's population has increased from 4,453 in 1990 and 6,162 in 2000 to 19,140 in 2010 (U.S. Census, 2002; U.S. Census accessed February 21, 2012, at *http://2010.census.gov/2010census/popmap/ipmtext.php?fl=13:1362104*). To satisfy its increasing water demand, Pooler began purchasing water from the City of Savannah in 1997. Pooler currently has two production wells (36Q283 and 36Q348; fig. 1) and is permitted to withdraw an annual daily average of 697,000 gallons per day (gal/d) from the Floridan aquifer system (Carol Couch, Director of the Georgia Environmental Protection Division, written commun., December 30, 2008).

Hydrogeologic Setting

Chatham County (fig. 1) is underlain by Coastal Plain strata consisting of consolidated to unconsolidated layers of sand and clay and semiconsolidated to dense layers of limestone and dolomite (Miller, 1986; Clarke and others, 1990; Williams and Gill, 2010). These sediments compose three major aquifer systems, in order of descending depth: the surficial aquifer system, the Brunswick aquifer system, and the Floridan aquifer system (fig. 2). Within the vicinity of Pooler, near Savannah, the Brunswick aquifer system has low permeability, with no discernible water-bearing units.

In the coastal area, the surficial aquifer system (fig. 2) consists of Miocene and younger interlayered sand, clay, and thin limestone beds (Clarke, 2003). Near the eastern boundary of the Hunter Army Airfield, about 8 mi southeast of the Pooler test site, the surficial aquifer system is about 100 ft thick and consists of an unconfined to semiconfined upper sand at a depth of 11–20 ft below land surface, and a confined lower sand at a depth of 37–56 ft below land surface. Well 36Q397, completed in the lower confined sand of the surficial aquifer system at the eastern boundary of Hunter Army Airfield, was pumped at a rate of 50 gallons per minute (gal/min) in 2011 (Gonthier, 2012). At Hunter Army Airfield, a confining unit consisting of silty clay and dense, phosphatic Miocene limestone (Clarke and others, 2011) transitions the base of the surficial aquifer system into the underlying Brunswick aquifer system.

The Brunswick aquifer system (fig. 2) consists of two water-bearing zones in the Brunswick, Ga. area—the upper Brunswick aquifer and the lower Brunswick aquifer (Clarke, 2003). The upper Brunswick aquifer consists of poorly sorted, fine to coarse, slightly phosphatic and dolomitic quartz sand, and dense phosphatic limestone (Clarke and others, 1990). The lower Brunswick aquifer consists of upper Oligocene and lower Miocene poorly sorted, fine to coarse, phosphatic and dolomitic sand (Clarke and others, 1990). The productivity of the Brunswick aquifer system is greatest in the vicinity of the Southeast Georgia embayment (Clarke, 2003), which is centered around St. Marys, Ga., near the Atlantic Coast, just north of the Georgia–Florida border (Miller, 1986; fig. 1). In Chatham County, the lower Brunswick aquifer is about 16 ft thick (Weems and Edwards, 2001), compared to 121 ft thick in the City of Brunswick. At well 36Q397 at the eastern boundary of Hunter Army Airfield, the Brunswick aquifer system is about 200 ft thick and consists of mostly fine-grain material having no major water-bearing units (Gonthier, 2012).

The principal source of water for all uses (excluding thermoelectric) in the coastal area of Georgia is the Floridan aquifer system (fig. 2). Williams and Gill (2010) noted that the Floridan aquifer system consists of a fairly thick sequence of carbonate rocks of mostly upper and middle Eocene age. The Floridan aquifer system is about 800 ft thick at the Berwick Plantation well (36Q330), about 3.73 mi southeast of the Pooler test site (Williams and Gill, 2010). The Floridan aquifer system is overlain by the silty clay and dense phosphatic lower

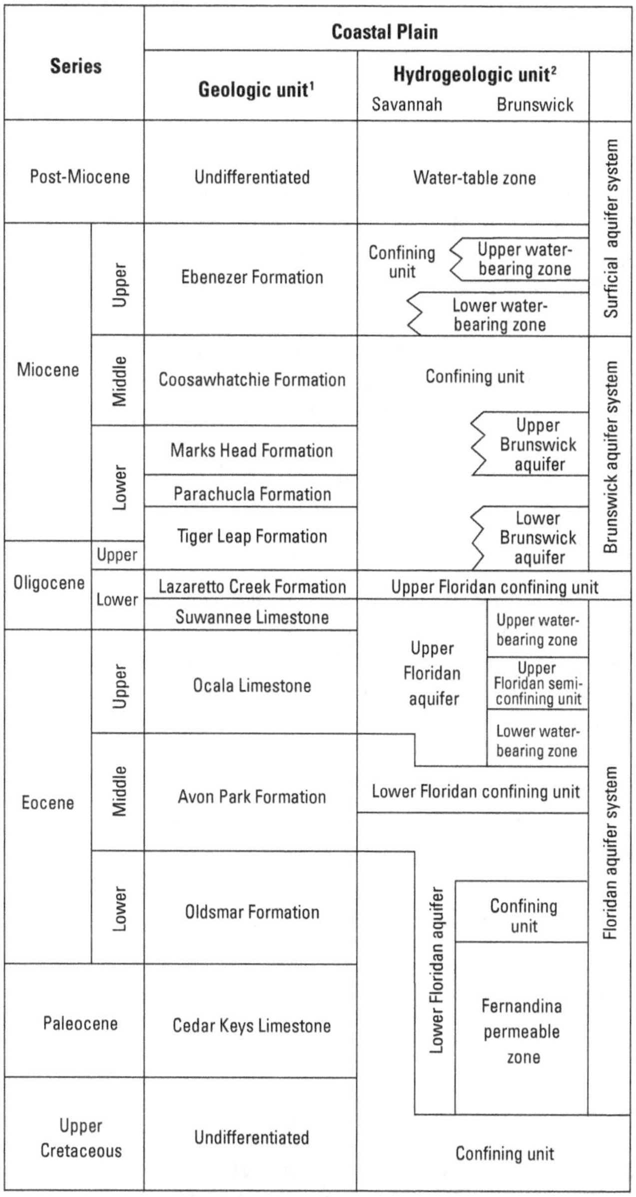

Series		Coastal Plain		
		Geologic unit[1]	Hydrogeologic unit[2] Savannah / Brunswick	
Post-Miocene		Undifferentiated	Water-table zone	Surficial aquifer system
Miocene	Upper	Ebenezer Formation	Confining unit / Upper water-bearing zone; Lower water-bearing zone	Brunswick aquifer system
	Middle	Coosawhatchie Formation	Confining unit	
	Lower	Marks Head Formation	Upper Brunswick aquifer	
		Parachucla Formation		
		Tiger Leap Formation	Lower Brunswick aquifer	
Oligocene	Upper	Lazaretto Creek Formation	Upper Floridan confining unit	Floridan aquifer system
	Lower	Suwannee Limestone		
Eocene	Upper	Ocala Limestone	Upper Floridan aquifer — Upper water-bearing zone; Upper Floridan semi-confining unit; Lower water-bearing zone	
	Middle	Avon Park Formation	Lower Floridan confining unit	
	Lower	Oldsmar Formation	Lower Floridan aquifer — Confining unit; Fernandina permeable zone	
Paleocene		Cedar Keys Limestone		
Upper Cretaceous		Undifferentiated	Confining unit	

[1] Modified from Randolph and others, 1991; Weems and Edwards, 2001
[2] Modified from Randolph and others, 1991; Clarke and Krause, 2000

Figure 2. Generalized correlation of geologic and hydrogeologic units in the Coastal Plain of Georgia.

Oligocene dolomite of the "upper confining unit" that separates the aquifer system from the overlying permeable units of the Brunswick aquifer system. Reported K_v of the upper confining unit, based on laboratory analysis of core, ranges from 2.3×10^{-4} to 3 feet per day (ft/d; Clarke and others, 2004) with two values reported for Chatham County of 8×10^{-4} and 4×10^{-3} ft/d (Furlow, 1969). The base of the Floridan aquifer system is underlain by low-permeability carbonate and clastic rocks of the "lower confining unit" (Williams and Gill, 2010).

The UFA (fig. 2) is highly productive and consists of upper Eocene to lower Oligocene limestone and dolomite. Williams and Gill (2010) reported an aquifer thickness in western Chatham County near Pooler, Ga., between 200 and 250 ft. Reported transmissivity of the UFA in Chatham County ranges from 20,000 to 80,000 feet squared per day (ft²/d; Clarke and others, 2004). Zones of high hydraulic conductivity exist within relatively thin intervals of the Floridan aquifer system, especially in the UFA (Clarke and others, 2004).

The UFA is underlain by the LFCU, consisting of dense, recrystallized middle Eocene limestone and dolomitic limestone that hydraulically separates, to varying degrees, the UFA from the LFA (fig. 2). Counts and Donsky (1963) reported that the K_v of this confining unit was 6.7×10^{-4} ft/d on the basis of laboratory analysis of a single core from Chatham County. The position and thickness of the LFCU was recently remapped in the area on the basis of flowmeter surveys and borehole geophysical logs (Williams and Gill, 2010). The LFCU is between 150 to 200 ft thick in western Chatham County in the vicinity of the Pooler test site and lies at an altitude between about −520 and −680 ft NAVD 88.

The LFA consists of middle Eocene limestone and dolomitic limestone but can include Paleocene and Upper Cretaceous carbonate units. As with the UFA, the LFA consists of several permeable water-bearing zones that are separated by layers of dense carbonate deposits. In western Chatham County in the vicinity of the Pooler test site, the LFA exists at altitudes ranging from about −680 to −1,120 ft NAVD 88, and has a reported thickness of about 450 ft (Williams and Gill, 2010). Transmissivity of the LFA at Berwick Plantation well 36Q330, 3.73 mi southeast of the Pooler test site, in Chatham County, Ga., is 8,200 ft²/d, (Clarke and others, 2004). Well 36Q392 completed in the LFA at Hunter Army Airfield, 7 mi southeast of the Pooler test site in Chatham County, has a transmissivity of 11,000 ft²/d (Williams, 2010).

The LFA is underlain by a lower confining unit (not to be confused with the LFCU), which consists of lower Eocene marl of low permeability (Williams and Gill, 2010; fig. 2). Falls and others (2005) describe the lithology of the lower confining unit as a semi-indurated, fine-grained mixture of carbonate, clay, silt, and sand that generally is dominated by clay and silt. In parts of the coastal area, the base of the Floridan aquifer system and the underlying marl is recognized on natural-gamma logs by a sharp increase in counts per second (Falls and others, 2005).

Well Identification

Wells in this report are identified by a USGS numbering system based on the index of USGS topographic maps (such as 35Q069). In Georgia, each 7-1/2-minute topographic quadrangle map has been given a number and letter designation beginning at the southwestern corner of the State. Numbers increase eastward through 39, and letters increase alphabetically northward through "Z" and then become double-letter designations "AA" through "PP." The letters "I" and "O" are not used. Wells inventoried in each quadrangle are numbered sequentially beginning with "1." For example, well 35Q069 is the 69th well inventoried in the Meldrim SE quadrangle (map 35Q).

Hydrogeology and Water Quality of the Floridan Aquifer System

To assess the hydrogeology and water quality of the Floridan aquifer system at the Pooler test site, multidiscipline site investigations were performed during 2011–2012 to collect and analyze geologic, geophysical, hydrologic, meteorological, and water-chemistry data. Analysis of these data provided a basis for refining the depth, thickness, hydraulic properties, and water quality of hydrogeologic units that compose the Floridan aquifer system in Chatham County, Ga.

Methods of Data Collection and Analysis

Hydrogeology and water quality of the Floridan aquifer system at the Pooler test site were assessed by installing two wells and performing geophysical logging, flowmeter surveys, water-quality sample collection and analyses, collection and hydraulic analysis of cores, packer-isolated and open slug tests, a 24-hour aquifer test of the UFA, and a 72-hour aquifer test of the LFA. Well installation included drilling a 1,131-ft-deep test hole followed by constructing a new 1,120-ft-deep well (35Q069) completed in the LFA (fig. 1). An observation well was constructed in the UFA (35Q070) 85 ft north of the new LFA well. Well construction information for all wells used during this study is listed in table 1. Data collection in

the new test wells included borehole geophysical logging, flowmeter surveys, water-quality sampling and analysis, core hydraulic analysis, packer-isolated slug tests, and aquifer tests.

Test Drilling and Well Installation

Mud-rotary drilling extended a borehole 302 ft into the sediments above the Floridan aquifer system at the Pooler test site. Drilling extended the borehole to 339 ft below land surface, just below the top of the Floridan aquifer system (figs. 3 and 4). A 24-inch-diameter surface casing was installed into the open hole to a depth of 80 ft; a 16-inch-diameter casing was installed to a depth of 339 ft.

Air-rotary drilling extended the borehole to a depth of 1,131 ft. An 8-inch-diameter casing was installed in the 318–705 ft depth interval, and the well (35Q069) was completed as an open hole to the LFA within the depth interval of 705–1,120 ft (fig. 4). The bottom 11 ft of the original 1,131 ft most likely was backfilled with native material.

The UFA well (35Q070) was drilled at a location 85 ft to the north of well 35Q069. Following installation of a 14-inch-diameter casing to a depth of 80 ft, mud-rotary drilling extended the borehole to a depth of 340 ft, just below the top of the Floridan aquifer system (fig. 5). A 6-inch-diameter casing was installed to 340 ft below land surface. Air-rotary drilling completed the boring into the UFA to a depth of 486 ft. The well was left as an open hole to the UFA from 340 to 486 ft below land surface.

Table 1.   Location and open intervals of wells used in the slug-test and aquifer-test analyses, Pooler, Georgia, 2011–2012, and wells from previous studies.

[Site locations, fig. 1; USGS, U.S. Geological Survey; NAVD 88, North American Vertical Datum of 1988; aquifer: FAS, Floridan aquifer system; UFA, Upper Floridan aquifer; LFA, Lower Floridan aquifer; SAS, surficial aquifer system; well 35Q069 was a test hole prior to being completed in the LFA; —, no data]

| USGS identifier | USGS site number | Land-surface altitude (feet above NAVD 88) | Depth (feet below land surface) | | | Aquifer |
			Static water level March 26, 2012	Top of open interval	Bottom of open interval	
35Q069	320356081162001	19	—	339	1,131	FAS
			41.99	705	1,120	LFA
35Q070	320357081162001	19	41.51	340	486	UFA
36Q283	320656081145801	23	—	280	610	UFA
36Q348	320650081144001	24	—	250	600	UFA
33R045	320754081364301	85	17.82	745	994	LFA
35P110	315443081185902	9.52	–19.62	315	441.25	UFA
35P125	315443081185903	11	–17.91	1,010	1,095	LFA
36Q020	320021081124801	13	–31.02	330	336	UFA
33P028	315434081364701	81.76	—	895	1,255	LFA
33P029	315434081364702	81.36	—	460	560	UFA
33P025	315447081345601	89	—	420	520	UFA
36Q330	320139081134002	11	—	718	1,080	LFA
36Q392	320005081102101	20	—	703	1,112	LFA
36Q397	320146081073701	24	—	35	70	SAS

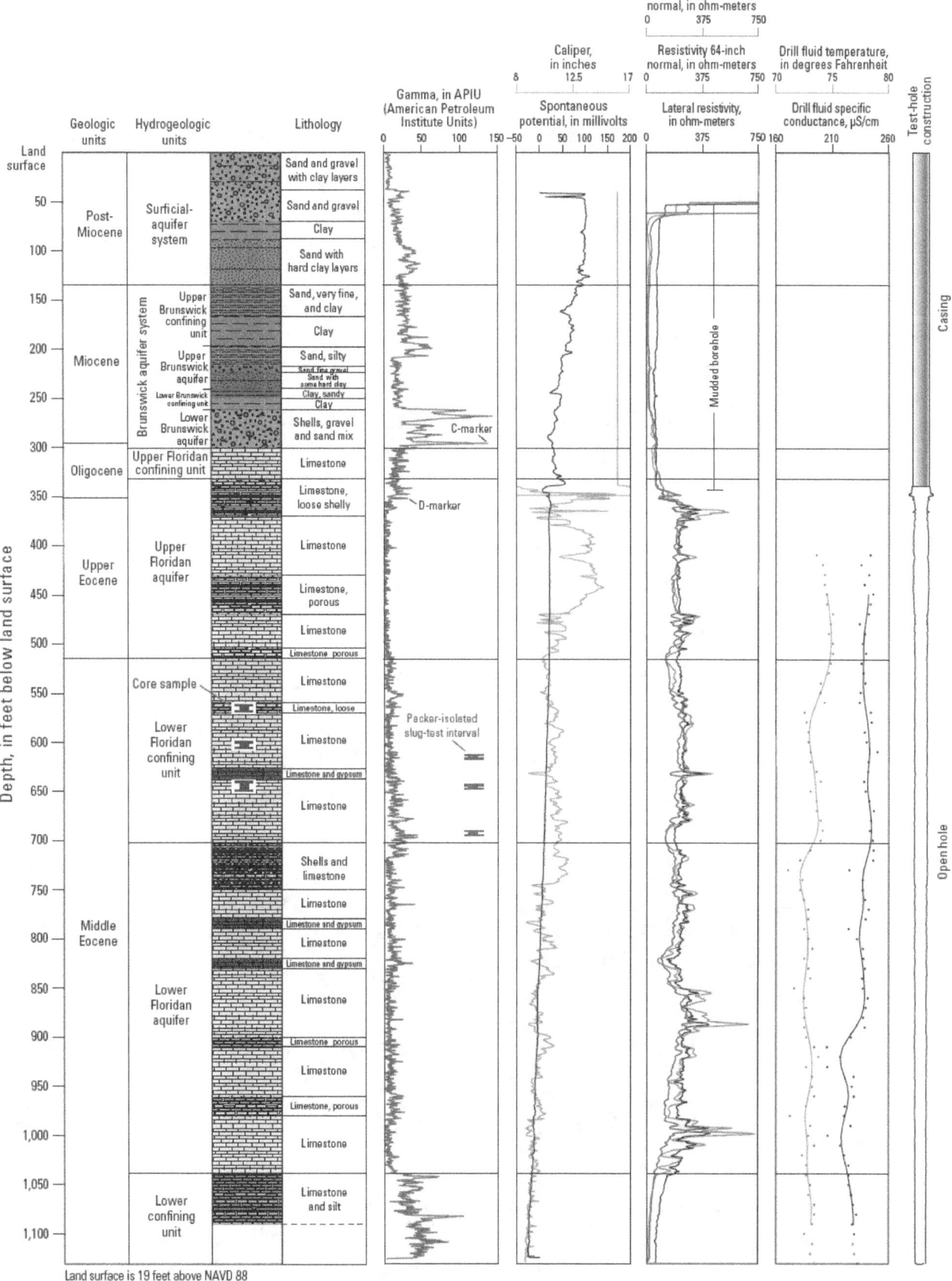

Figure 3. Lithology and geophysical properties of well 35Q069, Pooler, Georgia. [c.u., confining unit; μS/cm, microsiemens per centimeter at 25 degrees Celsius]

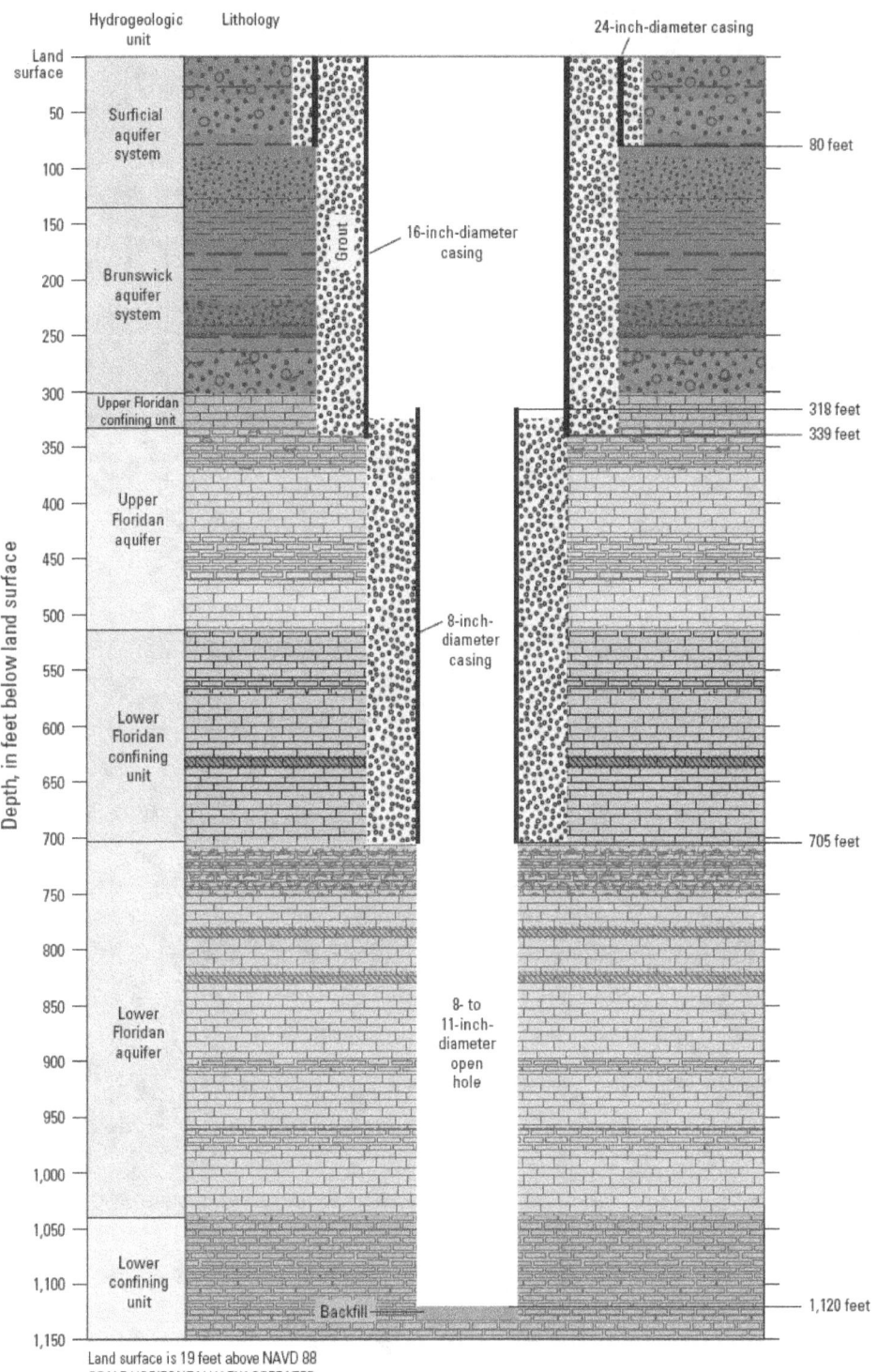

Figure 4. Hydrogeologic units and well completion diagram of Lower Floridan aquifer well 35Q069, Pooler, Georgia. Lithology patterns are labeled in figures 3 and 7.

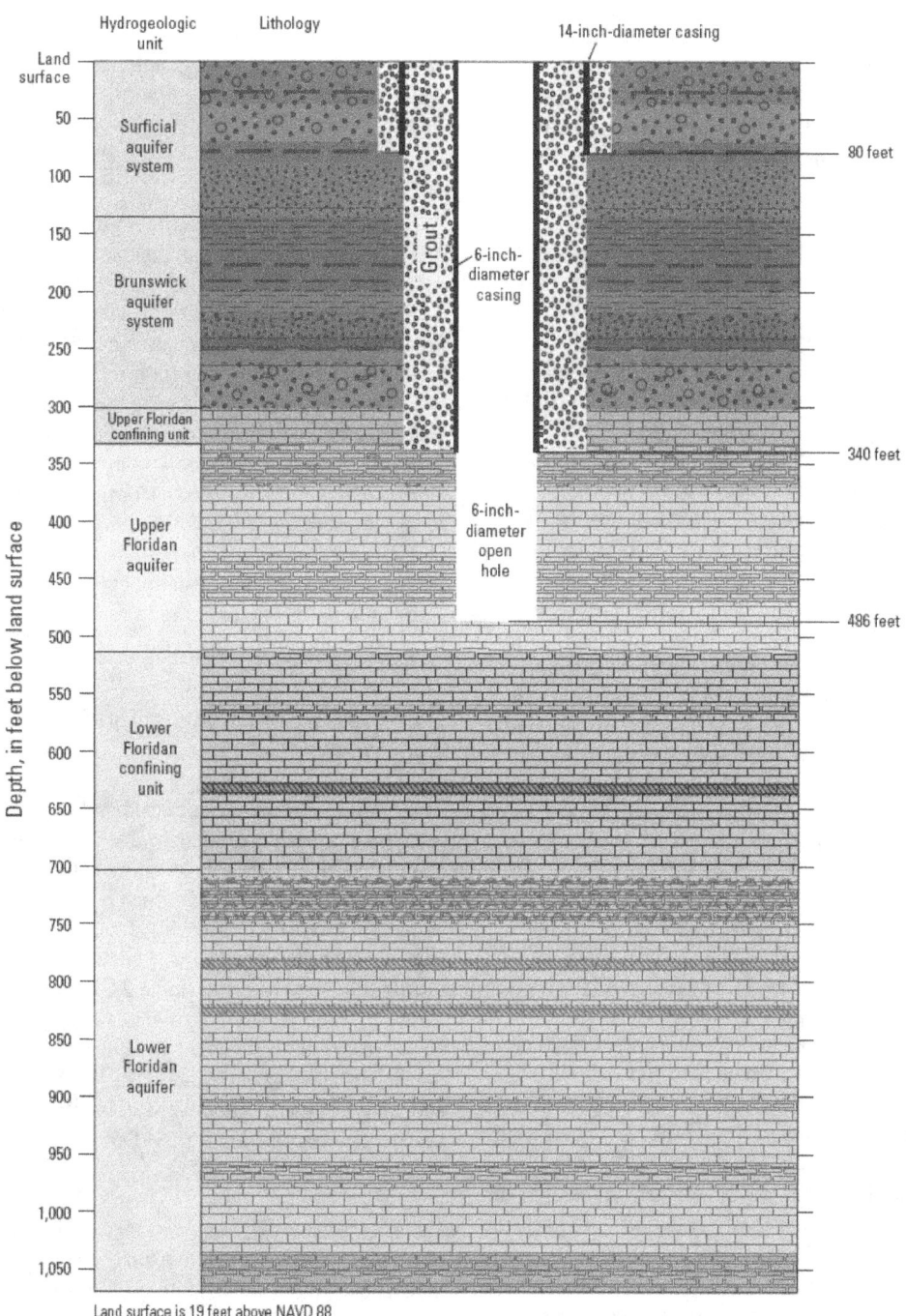

Figure 5. Hydrogeologic units and well completion diagram of Upper Floridan aquifer well 35Q070, Pooler, Georgia. Lithology patterns are labeled in figures 3 and 7.

Lithologic and Borehole Geophysical Logs

Drill cuttings collected every 10 ft from test hole 35Q069 were identified for grain size and mineral content; within the carbonate sequence, the amount of induration (cementation between grains) and the percentage of shells and recognized shell bits were also recorded. Although it is likely that the carbonate sequence contains dolostone, cuttings were not specifically tested for the presence of dolomite, and so the carbonate sequence is referred to as "limestone" in this report. In addition to drill cuttings, limestone cores were extracted and examined for the depth intervals of 564.3–569.8, 600–605, and 640–648.4 ft (fig. 3). Secondary porosity in the form of fractures and dissolution features present within the core was not discernible in drill cuttings from the same depth intervals. Drilling fluid also was sampled every 10 ft for specific conductance and temperature.

Borehole-geophysical logs were collected at various stages of drilling well 35Q069 to characterize the physical properties of penetrated sediments, rock, and interstitial fluid. The first set of logs was collected in the 0–339 ft interval where mud-rotary drilling penetrated clastic sediments overlying the Floridan aquifer system. The second set of logs was collected in the 339–1,131 ft interval where air-rotary drilling was used to penetrate the carbonates of the Floridan aquifer system. In both intervals, the following logs were collected: caliper (shown on figure 3 as the inner diameter of the casing from 0–339 ft); natural gamma; spontaneous potential, and single-point lateral, long- and short-normal resistivity. In the deeper carbonate interval, borehole-fluid resistivity and temperature were collected.

Flowmeter Survey

Flowmeter surveys were performed to quantify the relative contributions of flow from water-bearing zones within the Floridan aquifer system including confinement between the UFA and LFA (fig. 6). Two flowmeter surveys were performed in well 35Q069 at two different stages of the well's construction. The first flowmeter survey was performed when the well was still a test hole open to the entire Floridan aquifer system. The second flowmeter survey was performed after the well was complete and was open only to the LFA. To perform a flowmeter survey, the well was pumped while several traverses were made in the open borehole with an electromagnetic (EM) flowmeter that measured accumulated flow up the borehole. The first flowmeter survey pumped the test hole for well 35Q069 at a rate of 708 gal/min on December 8, 2011. Information from this flowmeter survey (1) ensured accurate placement of the well casing below the LFCU during completion of the LFA well, and (2) was used to calculate the concentrations of constituents in water from intervals between composite grab samples. The second flowmeter survey pumped the completed LFA well 35Q069 at a rate of about

783 gal/min on April 18, 2012, in conjunction with a 72-hour aquifer test. The second flowmeter survey did not affect the results of the 72-hour aquifer test. The water-level hydrograph of the pumped well during the flowmeter survey was inspected. There was no indication that the flowmeter survey created water-level fluctuations during the aquifer test.

Water-Level Measurements

Continuous and intermittent groundwater-level measurements were made at the borehole according to USGS standard procedures (Garber and Koopman, 1968; Stallman, 1971; and Cunningham and Schalk, 2011). Manual intermittent water-level measurements were made for calibration of groundwater-level recorder readings and for direct monitoring of background water-level trends during the performance of the slug tests. Manual measurements were made to the nearest 0.01 ft using an electric tape and following procedures described in Garber and Koopman (1968). Continuous groundwater-level recorders were equipped with submerged, vented pressure transducers. For long-term monitoring, pressure transducers collected water-level measurements every 15 minutes; for the slug-test monitoring, pressure transducers collected water-level measurements every second to every 15 minutes.

Core Hydraulic Analysis and Packer-Isolated Slug Tests

Core samples were collected and analyzed for K_v and porosity at a testing laboratory. Packer-isolated slug tests were completed in the borehole to estimate horizontal hydraulic properties. Relatively undisturbed core samples were collected at depth intervals of 564.3–569.8, 600–605, and 640–648.4 ft and submitted to S&ME, Inc. (Atlanta, Ga.) for hydraulic testing of K_v and porosity (table 2). To retain the undisturbed nature of these largely consolidated core samples, the samples were preserved onsite using procedures described in American Society for Testing and Materials (ASTM) D5079 (2008) and analyzed using a flexible wall permeameter following procedures described in ASTM D5084 (2010).

Table 2. Estimated vertical hydraulic conductivity and porosity of core samples collected from the Lower Floridan confining unit in test hole 35Q069, Pooler, Georgia, November 2011.

[Analyses by S&ME, Inc., Atlanta, Georgia; ft/d, foot per day]

Core sample interval (feet below land surface)	Vertical hydraulic conductivity (ft/d)	Porosity
564.3–569.8	1.67	0.34
600–605	1.08	0.33
640–648.4	0.57	0.33

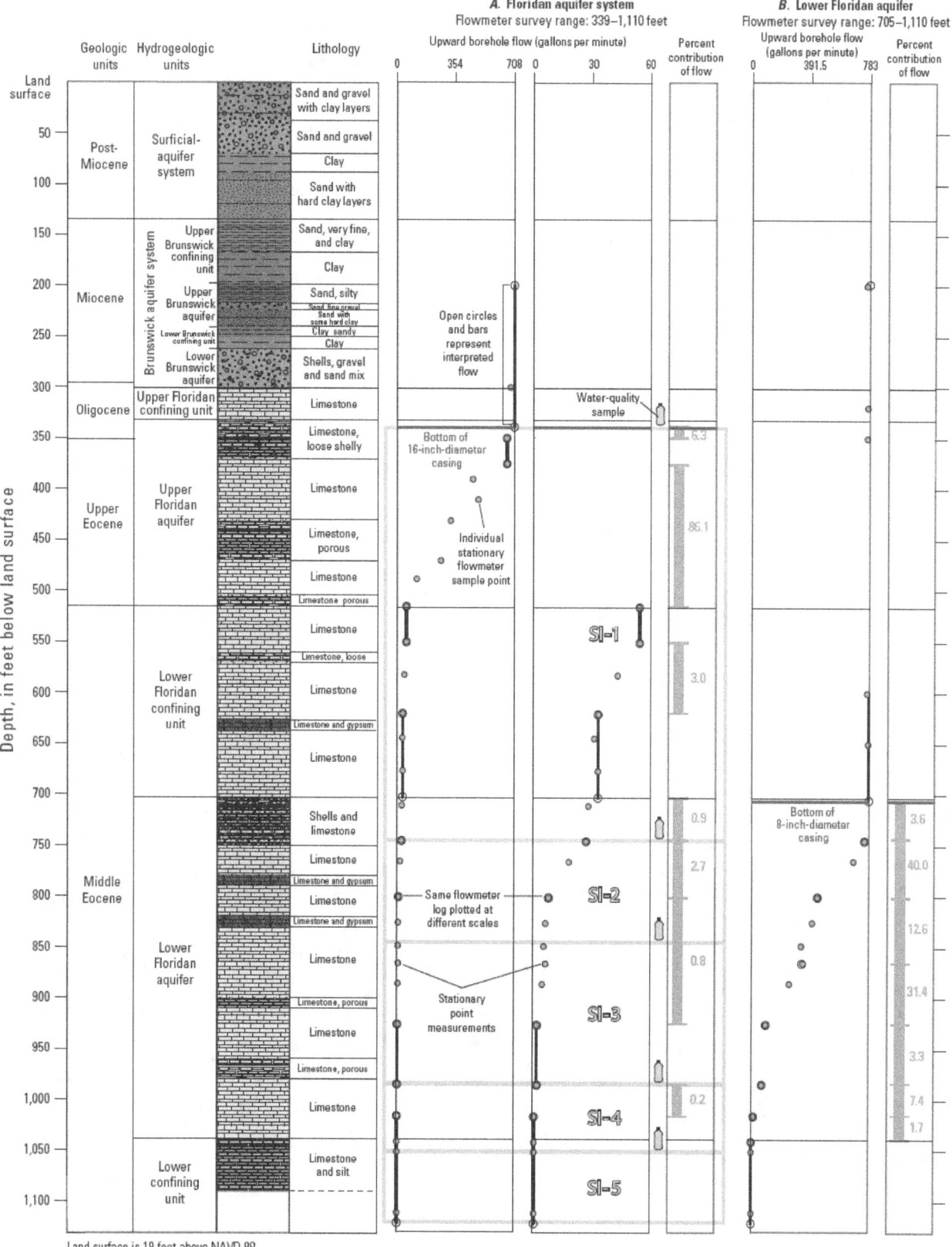

Figure 6. Water-quality-sample collection locations in test hole 35Q069 and flowmeter-survey data from *(A)* pumping the Floridan aquifer system prior to installing an 8-inch-diameter casing in test hole 35Q069, and *(B)* pumping the Lower Floridan aquifer after installing an 8-inch-diameter casing in test hole 35Q069 and completing it as well 35Q069, Pooler, Georgia. [SI, sample interval between water-quality samples]

Packer-isolated slug tests at selected intervals—612–618 ft, 642–648 ft, and 690–696 ft (fig. 3)—during December 13–14, 2011, were performed in the test hole for well 35Q069 to obtain a sample of values of K_h of the LFCU (table 3). Each test interval was isolated using a straddle packer (fig. 7). The packer assembly sealed off a 6.05-ft depth interval of the aquifer and hydraulically connected the aquifer to a 3.45-inch-inner-diameter casing. A slug was inserted (falling-head test) or removed (rising-head test) from the water column within this connected casing. The slug consisted of a 10.17-ft-long, 2.37-inch-diameter polyvinyl-chloride pipe filled with sand and capped on each end. The volume of the slug totaled 0.31 cubic foot (ft^3) and displaced 4.8 ft of the water column, which fluctuated within the 3.45-inch-inner-diameter casing during the test. The change in head (water level) with time was recorded during the tests.

Pressure transducers monitored water-level response within the isolated interval as well as above and below the zones being tested. The lower transducer was connected to the packer assembly just below the lower packer, 10.48 ft below the middle transducer. The upper transducer was lowered into the borehole, roughly 150 ft below land surface, within the part of the borehole that was cased off from the Miocene to post-Miocene sediments. Both packers in the packer assembly were connected to the same air line for inflation. The general procedures for conducting the packer-isolated slug tests were as follows:

1. Lower packer assembly to the target depth.

2. Attach pressure transducers to dataloggers and start recording at 1-second intervals.

3. Inflate both packers to a specific pressure.

4. Retrieve and review water-level data to confirm reestablishment of static conditions after packer inflation.

5. Perform falling-head slug test by submerging slug into water column connected to packer-isolated interval to raise hydraulic head within the test interval.

6. Retrieve and review water-level data to confirm return to pre-test conditions and to assess any leakage beyond the packer-isolated interval.

7. Increase pressure to further inflate packers.

8. Retrieve and review water-level data to confirm reestablishment of static conditions after packer inflation.

9. Perform rising-head slug test by removing slug from water column connected to packer-isolated interval to lower hydraulic head within the test interval.

10. Retrieve and review water-level data to confirm return to pre-test conditions and to assess any leakage beyond the packer-isolated interval.

Table 3.   Hydraulic conductivity values for depth intervals emphasizing the Lower Floridan confining unit in test hole 35Q069, Pooler, Georgia, December 13–14, 2011.

[ft/d, feet per day; ft²/d, foot squared per day; hydrogeologic unit: LFCU, Lower Floridan confining unit; FAS, Floridan aquifer system; methods by Bouwer and Rice (1976) or van der Kamp (1976) are contained in spreadsheets from Halford and Kuniansky (2002); —, not applicable]

Depth interval (feet below land surface)	Hydrogeologic unit	Type	Month and day	Time	Analysis method	Horizontal hydraulic conductivity (ft/d)	Transmissivity (ft²/d)[a]
612–618	LFCU	Falling head	December 13	3:29:26 p.m.	Bouwer and Rice	10	—
612–618	LFCU	Rising head	December 13	3:39:32 p.m.	Bouwer and Rice	10	—
642–648	LFCU	Falling head	December 14	8:46:48 a.m.	Bouwer and Rice	3	—
642–648	LFCU	Rising head	December 14	9:01:08 a.m.	Bouwer and Rice	3	—
690–696	LFCU	Rising head	December 14	12:08:45 p.m.	Bouwer and Rice	0.5	—
339–1,130	FAS	Falling head	December 14	1:11:42 p.m.	van der Kamp	—	54,000
339–1,130	FAS	Rising head	December 14	1:17:58 p.m.	van der Kamp	—	62,000

[a]A storage coefficient of 1.7×10^{-3} was used for the Floridan aquifer system.

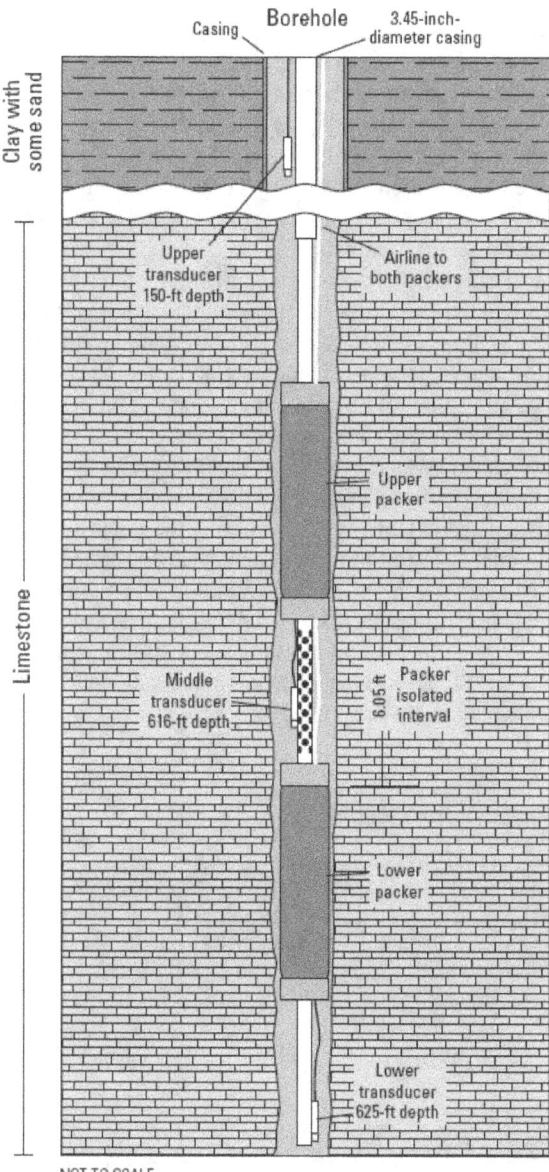

Figure 7. Straddle-packer assembly in test hole 35Q069 as an example at the 612- to 618-feet packer-isolated interval, Pooler, Georgia, December 13, 2011.

The caliper log was used to position the packers in smooth sections of the borehole for good packer seating. The flowmeter survey and geophysical logs were used to locate parts of the borehole (penetrating the LFCU) that were especially tight or nonpermeable (fig. 3); these nonpermeable parts of the LFCU were targeted for slug tests. Packer pressure was increased from one test to the next at a given depth to better detect any possible leakage within the packers. A decrease in recovery rate with increased pressure to inflate packers would indicate the presence of leakage.

The K_h of the surrounding material of the LFCU was calculated using water-level-recovery data following a slug test and the Bouwer and Rice method (Bouwer and Rice, 1976), which was contained in the spreadsheet developed by Halford and Kuniansky (2002). The log of the fraction of remaining water-level displacement to the initial displacement was plotted on the y-axis, and corresponding time was plotted on the x-axis. This semi-log plot depicts the exponential decay of the fractional displacement in water level, or water-level recovery, with time as a straight line. The spreadsheet by Halford and Kuniansky (2002) utilizes two analytical points on the water-level recovery line in an analysis plot. The slope of these two aligned points and the geometries of the well opening and aquifer are then used to estimate the K_h.

In addition to slug tests performed on packer-isolated intervals, falling-head and rising-head slug tests were performed in the test hole when it was open to the entire Floridan aquifer system prior to completion as well 35Q069. Water levels in the well exhibited underdamped responses (oscillations) during the slug tests. Transmissivity of the Floridan aquifer system was calculated using the underdamped water-level-recovery data obtained from the two slug tests, and the van der Kamp method (van der Kamp, 1976), which was contained in the spreadsheet developed by Halford and Kuniansky (2002). This spreadsheet allows positioning of analytical points on an analysis plot to define the frequency and damping rate of the oscillations. The frequency and damping rate of the oscillations, test-hole diameter, and a value of the coefficient of storage are used to estimate the transmissivity (table 3). Details and results of the slug tests are discussed in appendix 1.

Aquifer Tests

Aquifer tests performed at the Pooler test site estimated the transmissivity of the Upper and Lower Floridan aquifers and quantified the effects of pumping one aquifer on water levels in the other aquifer (table 4 and fig. 8). Water levels were monitored in wells 35Q069 and 35Q070 during both aquifer tests (fig. 9). A 24-hour aquifer test was performed during March 27–28, 2012, in well 35Q070 open to the UFA, and a 72-hour aquifer test was performed during April 16–19, 2012, in well 35Q069 open to the LFA.

Drawdown response to the 24-hour aquifer test was analyzed using the Cooper and Jacob (1946) method to determine the transmissivity of the UFA. Drawdown response to the 72-hour aquifer test was simulated using MODOPTIM (Halford, 2006b) to determine the transmissivity of the UFA, LFCU, and LFA, as well as the storage coefficient of the UFA and the LFCU. Aquifer test analyses and results are discussed in detail in appendix 2.

Drawdown Estimation

Drawdown for the two monitored wells was estimated by using a tool developed by Halford (2006a) to filter water-level data for effects of barometric pressure, earth tide, ocean tide, and long-term trends. The filtering tool has a time-series spreadsheet that is used to match (fit) synthetic water levels to the measured water levels during a period that is unaffected by an aquifer test or local pumping event (hereinafter referred to as the "fitting period"). Data and parameters that were used to fit the synthetic water levels to measured water levels are then used to fit synthetic water levels to measured water levels before the start of aquifer-test pumping and after the end of aquifer-test recovery. With the synthetic water level being the same as the measured water level at the start of the aquifer test, the synthetic water level minus the measured water level during the aquifer test yielded the drawdown (maximum drawdown listed in table 4). After filtering the water-level data for nonaquifer-test influences, the data were culled to be spaced evenly with log (time) to minimize the number of data points used to match simulated drawdown to measured drawdown. Water level used to estimate drawdown is in feet above NAVD 88. Drawdown estimation methods and results are discussed in detail in appendix 2.

Table 4.   Summary of slug- and aquifer-test results, Pooler, Georgia, December 2011–April 2012.

[Hydrogeologic unit: FAS, Floridan aquifer system; UFA, Upper Floridan aquifer; LFCU, Lower Floridan confining unit; LFA, Lower Floridan aquifer; ft, foot; gal/min, gallon per minute; ft²/d, foot squared per day; —, not applicable]

	Hydrogeologic unit	Slug test	Aquifer test	
		Open hole	24-hour	72-hour
Test information				
Location	—	Test hole 35Q069	Well 35Q070	Well 35Q069
Interval tested (ft)	—	FAS	UFA	LFA
Test date	—	December 14, 2011	March 27–28, 2012	April 16–19, 2012
Pumping rate (gal/min)	—	—	285	783
Analysis	—	van der Kamp	Cooper-Jacob	MODOPTIM with MODFLOW
Maximum drawdown (feet)				
Well 35Q070	UFA	—	11.1	0.9
Well 35Q069	LFA	—	0.09	51.7
Coefficients				
Transmissivity (ft²/d)	FAS	58,000	—	51,000
Transmissivity (ft²/d)	UFA	—	30,000	46,000
Storage	UFA	—	—	5.90×10^{-4}
Transmissivity (ft²/d)	LFCU	—	—	700
Storage	LFCU	—	—	4.50×10^{-4}
Transmissivity (ft²/d)	LFA	—	—	4,000

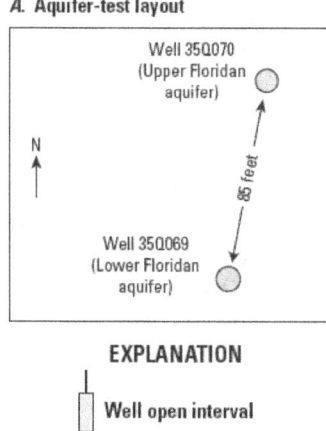

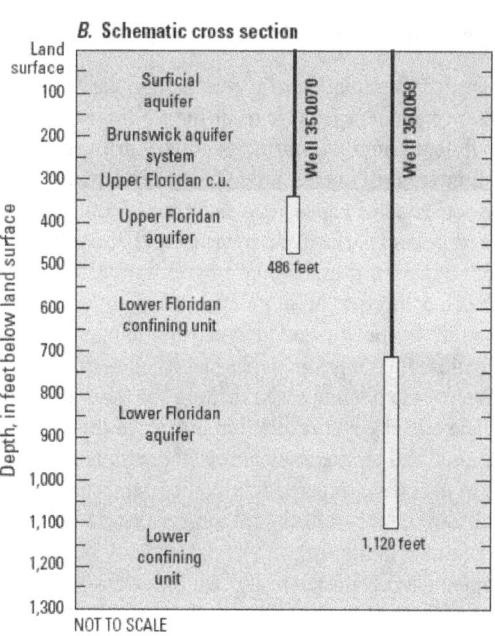

Figure 8.   Location and open interval of wells used for the 24-hour and 72-hour aquifer tests, Pooler, Georgia. *A*, Diagram showing aquifer-test layout. *B*, Schematic cross section showing major hydrogeologic units. [c.u., confining unit]

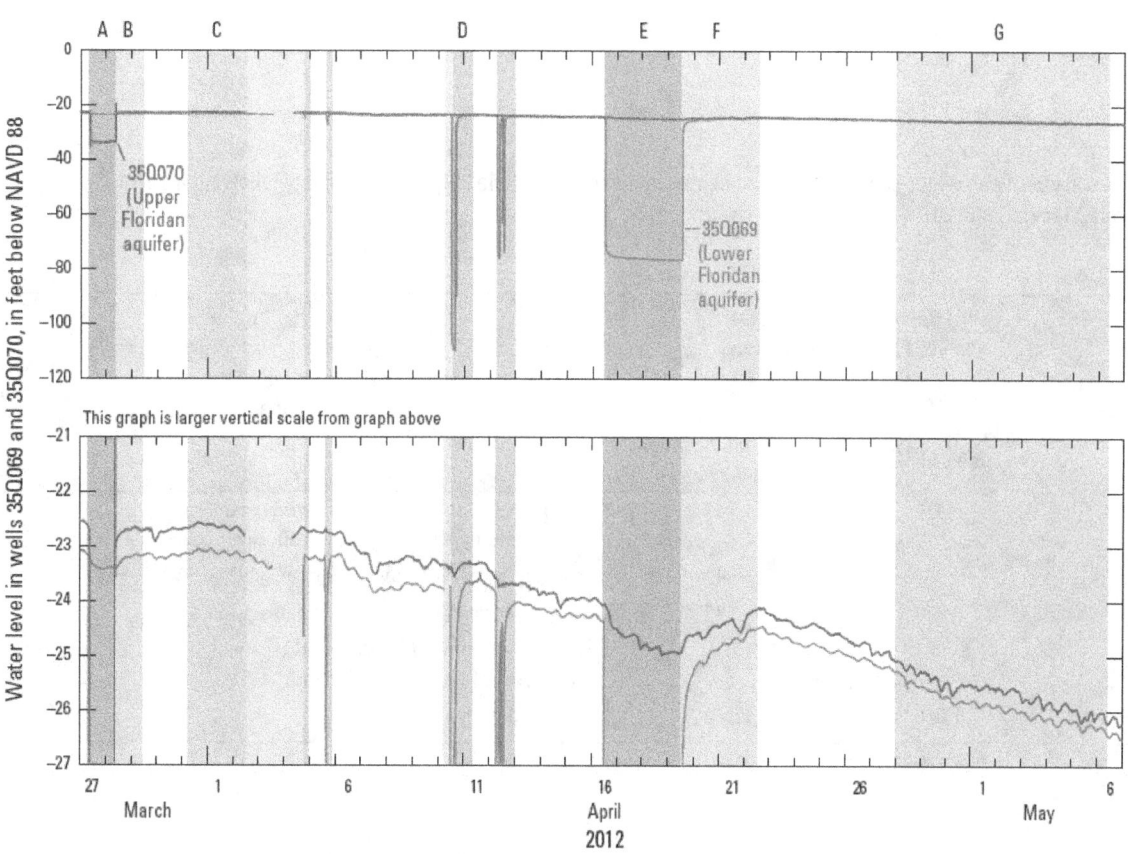

Figure 9.   Water levels in wells 35Q069 and 35Q070 before, during, and after the 24-hour Upper Floridan and 72-hour Lower Floridan aquifer tests, Pooler, Georgia, March 27–May 6, 2012. *A*, Pumping phase of Upper Floridan 24-hour aquifer test. *B*, Recovery phase of Upper Floridan 24-hour aquifer test. *C*, Early fitting period. *D*, Failed Lower Floridan aquifer test. *E*, Pumping phase of Lower Floridan 72-hour aquifer test. *F*, Recovery phase of Lower Floridan 72-hour aquifer test. *G*, Late fitting period. The fitting period is used to describe the match (or fit) of synthetic water levels to those that were measured.

EXPLANATION

Pumping phase of aquifer test

Recovery phase of aquifer test

Fitting period

Data gap in at least one well

Other pumping event

Water-Quality Sampling and Analysis

To assess vertical distribution of water quality, the specific conductance and temperature of drilling fluid were measured at 10-ft depth intervals during air-rotary drilling of the 360–1,131 ft interval in the test hole (fig. 3). The measurement procedure consisted of capturing a sample of drilling fluid as it emerged at land surface and measuring the specific conductance after every 10 ft of drilling progression. Although discharge water is a composite of all units exposed above a given depth, changes in specific conductance and temperature provide an indication of changes in water quality with depth in the test hole. Water was added to assist drilling the 360–370 ft depth interval. The added water resulted in lower values for specific conductance and temperature associated with the 360–400-ft depth interval compared to all other depths. As a result, the data for the 360–410-ft depth interval were not used and are not shown in figure 3.

Water samples were collected in the test hole for well 35Q069 on December 8, 2011, immediately following a borehole-flowmeter survey using a wireline grab sampler at five distinct depths (table 5 and fig. 6). The test hole was pumped for at least an hour during the borehole-flowmeter survey before water samples were collected. Sample-collection points were located at depths of 339, 745, 845, 985, and 1,050 ft. Water collected at each sample-collection point is a composite of all water entering the well below the sample-collection point. These grab samples were collected with the pump set within the casing above all water-bearing intervals and pumping at a rate of 708 gal/min; therefore, each of these samples represents a composite of the water entering the test hole beneath a specific sampling depth. Water was transferred from the grab sampler to sample bottles using a peristaltic pump. Samples were analyzed for specific conductance, pH, alkalinity reported as calcium carbonate, and total dissolved solids, in additional to major ions, including calcium, magnesium, sodium, potassium, iron, manganese, sulfate, chloride, and fluoride (table 6). Water collected for major ions was filtered using a capsule filter with a 0.45-micrometer (µm) pore medium. Samples for cations were preserved with nitric acid. Samples were analyzed at TestAmerica Laboratories, Inc., Savannah, Ga. Cations were analyzed using induced coupled plasma; anions were analyzed using ion chromatography. Bicarbonate concentrations were calculated from values of alkalinity.

Table 5.   Intervals between water-sample collection depths in pumped test hole 35Q069 immediately following a flowmeter survey, Pooler, Georgia, December 8, 2011.

[<, less than]

Sample interval (SI) number	Depth of sample interval (feet below land surface)		Hydrogeologic unit(s) to which the sample interval is open	Percent contribution of flow
	Top (also depth of grab sample)	Bottom (also depth of grab sample for the lower sample interval)		
SI-1	339	745	Upper Floridan aquifer; Lower Floridan confining unit; small, upper part of Lower Floridan aquifer	96.3
SI-2	745	845	Central-upper part of Lower Floridan aquifer	3.0
SI-3	845	985	Central-lower part of Lower Floridan aquifer	0.5
SI-4	985	1,050	Lower part of Lower Floridan aquifer; small, upper part of lower confining unit	0.2
SI-5	1,050	1,120	Lower confining unit	<0.01

Table 6. Water quality of composite samples and sample intervals immediately following a flowmeter survey in pumped test hole 35Q069, Pooler, Georgia, December 8, 2011.

[Range of fluoride concentrations caused by censored values from composite samples. Composite-sample results and flow contribution from screens were used to calculate water quality in sample intervals. Criteria type: SDWR, Secondary Drinking Water Regulation; HBVs, health-based value for individuals on a 500 milligrams per day (mg/d) restricted sodium diet; µS/cm, microsiemens per centimeter at 25 degrees Celsius; SU, pH standard units; mg/L, milligram per liter; µg/L, microgram per liter; —, no water-quality criteria for this constituent or did not calculate concentration because of censored values from composite samples; <, less than]

Constituent parameter	Units	Water-quality values of composite samples with depth range (feet below land surface)					[a]Water quality of individual sample intervals (SI) by depth range (feet below land surface)					Water-quality criteria[k]	Criteria type
		1,120–1,050[b]	1,120–985[c]	1,120–845[d]	1,120–745[e]	1,120–339[f]	1,120–1,050[b] SI-5	1,050–985[g] SI-4	985–845[h] SI-3	845–745[i] SI-2	745–339[j] SI-1		
Specific conductance	µS/cm	277	238	234	230	235	277	238	232	229	235	—	—
pH	SU	8.17	8.04	8.02	8.05	8.00	8.17	8.04	8.01	8.06	8.00	6.5–8.5	SDWR
Alkalinity	mg/L	134	113	110	119	108	134	113	109	121	108	—	—
Total dissolved solids	mg/L	204.0	182.0	184.0	168.0	199.0	204.0	182.0	184.8	164.2	200.2	500	SDWR
Calcium	mg/L	19.80	28.80	29.40	26.60	28.70	19.80	28.80	29.63	25.93	28.78	—	—
Magnesium	mg/L	5.85	8.22	8.22	8.63	7.56	5.85	8.22	8.22	8.73	7.52	—	—
Sodium	mg/L	30.50	10.90	10.10	9.05	8.50	30.50	10.90	9.80	8.80	8.48	20	HBVs
Potassium	mg/L	3.08	2.28	2.51	2.05	1.87	3.08	2.28	2.60	1.94	1.86	—	—
Iron	µg/L	153	<100	<100	<100	<100	153	—	—	—	—	300	SDWR
Manganese	µg/L	<10	<10	<10	<10	<10	<10	—	—	—	—	50	SDWR
Bicarbonate	mg/L	163	138	134	145	132	163	138	133	148	131	—	—
Sulfate	mg/L	11.90	5.97	5.78	5.71	5.37	11.90	5.97	5.71	5.69	5.36	500	HBV
Chloride	mg/L	8.05	5.22	5.26	5.09	5.07	8.05	5.22	5.28	5.05	5.07	250	SDWR
Fluoride	mg/L	0.50	0.36	0.34	0.33	0.30	0.50	0.36	0.33	0.33	0.30	2	SDWR
Contribution of flow	percent	<0.01	0.2	0.7	3.7	100.0	<0.01	0.2	0.5	3.0	96.3	—	—

[a]Water-quality values of composite samples and flowmeter-survey data (contribution of water-bearing zones) were used to calculate water-quality value of sample interval.

[b]Lower confining unit

[c]Lower confining unit and small, lower part of Lower Floridan aquifer.

[d]Lower confining unit and lower half of the Lower Floridan aquifer.

[e]Lower confining unit and most of the Lower Floridan aquifer.

[f]Lower confining unit and Floridan aquifer system.

[g]Small lower part of Lower Floridan aquifer.

[h]Central lower part of Lower Floridan aquifer.

[i]Central upper part of Lower Floridan aquifer.

[j]Small upper part of Lower Floridan aquifer, Lower Floridan confining unit, and Upper Floridan aquifer.

[k]U.S. Environmental Protection Agency, 2011.

A simple mixing equation and the flow contribution from water-bearing units from the borehole-flowmeter survey were used to convert composite water-sample concentrations into concentrations of individual sample intervals between sample-collection points (fig. 10). Water was assumed to be flowing from adjacent hydrogeologic units and completely mixing before reaching the collection point. The top four sample intervals represent the Floridan aquifer system (table 5). The UFA, LFCU, and the upper part of the LFA are represented by sample interval one or SI-1; the central-upper part of the LFA is represented by SI-2; the central-lower part of the LFA is represented by SI-3; and the lower part of the LFA associated with high resistivity values is represented by SI-4. The lower confining unit, underlying the Floridan aquifer system, is represented by SI-5. The mixing equation from Kendall and Caldwell (1998, p. 80) was applied to sample intervals in the test hole as follows:

$$Q_{T,n}C_{T,n} = Q_{T,n-1}C_{T,n-1} + Q_{I,n}C_{I,n}, \quad (1)$$

where

$Q_{T,n}$ is the composite discharge at sample-collection point n, contributed to or flowing up the borehole from all water-bearing intervals below sample-collection point n, in gallons per minute;

$C_{T,n}$ is the concentration of a specific conservative constituent in discharge water $Q_{T,n}$, expressed in a linear-unit value that varies with constituent, but represents the mass of the constituent per volume of water;

$Q_{T,n-1}$ is the composite discharge at sample-collection point $n-1$, contributed to or flowing up the borehole from all water-bearing intervals below sample-collection point $n-1$, in gallons per minute;

$C_{T,n-1}$ is the concentration of a specific conservative constituent in discharge water $Q_{T,n-1}$, expressed in a linear-unit value that varies with constituent, but represents the mass of the constituent per volume of water;

$Q_{I,n}$ is the discharge entering the well from the interval between sample-collection points n and $n-1$, in gallons per minute; and

$C_{I,n}$ is the concentration of a specific conservative constituent in discharge water $Q_{I,n}$, expressed in a linear-unit value that varies with constituent, but represents the mass of the constituent per volume of water.

Discharge rates are known from the borehole-flowmeter survey, and the composite water-sample concentrations

at sample-collection points are known from sampling and analysis. Therefore, equation 1 can be rearranged to solve for the concentration, $C_{I,n}$, of the specific conservative constituent in discharge water ($Q_{I,n}$) entering the well between the two sample-collection points n and $n-1$:

$$C_{I,n} = \frac{Q_{T,n}C_{T,n} - Q_{T,n-1}C_{T,n-1}}{Q_{I,n}}. \quad (2)$$

In addition to the grab water samples collected in the test borehole, a composite water sample was collected from completed LFA well 35Q069 after 71 hours of pumping during the 72-hour aquifer test on April 19, 2012. The water sample was analyzed for pH, total dissolved solids, color, alkalinity, dissolved carbon dioxide, turbidity, hardness as calcium carbonate, iron, manganese, zinc, sulfate, chloride, fluoride, and nitrate plus nitrite (table 7). Bicarbonate concentration was calculated from the value of alkalinity. Over 600 well volumes were pumped from the well prior to sampling water from the LFA. Samples were analyzed at TestAmerica Laboratories, Inc., Savannah, Ga. Water type for sample intervals from the test borehole and the composite sample for well 35Q069 was determined from the percentage of equivalents of sodium plus potassium, calcium, magnesium, chloride plus fluoride, sulfate, and carbonate plus bicarbonate that were plotted on a piper diagram (fig. 11; Piper, 1944).

Table 7. Water-quality results for a water sample collected from well 35Q069, Pooler, Georgia, April 19, 2012. Water sample was collected 71 hours into the 72-hour aquifer test.

[SU, standard pH units; mg/L, milligram per liter; PCU, platinum-cobalt unit; NTU, nephelometric turbidity unit; µg/L, microgram per liter; <, less than]

Constituent parameter	Value	Unit
pH	8.15	SU
Total dissolved solids	190	mg/L
Color	<5.0	PCU
Alkalinity	110	mg/L
Dissolved carbon dioxide	<5.0	mg/L
Turbidity	0.13	NTU
Hardness as calcium carbonate	81	mg/L
Iron	<50	µg/L
Manganese	<10	µg/L
Zinc	<20	µg/L
Bicarbonate	134	mg/L
Sulfate	14	mg/L
Chloride	6.4	mg/L
Fluoride	0.68	mg/L
Nitrate plus nitrite as nitrogen	<0.05	mg/L

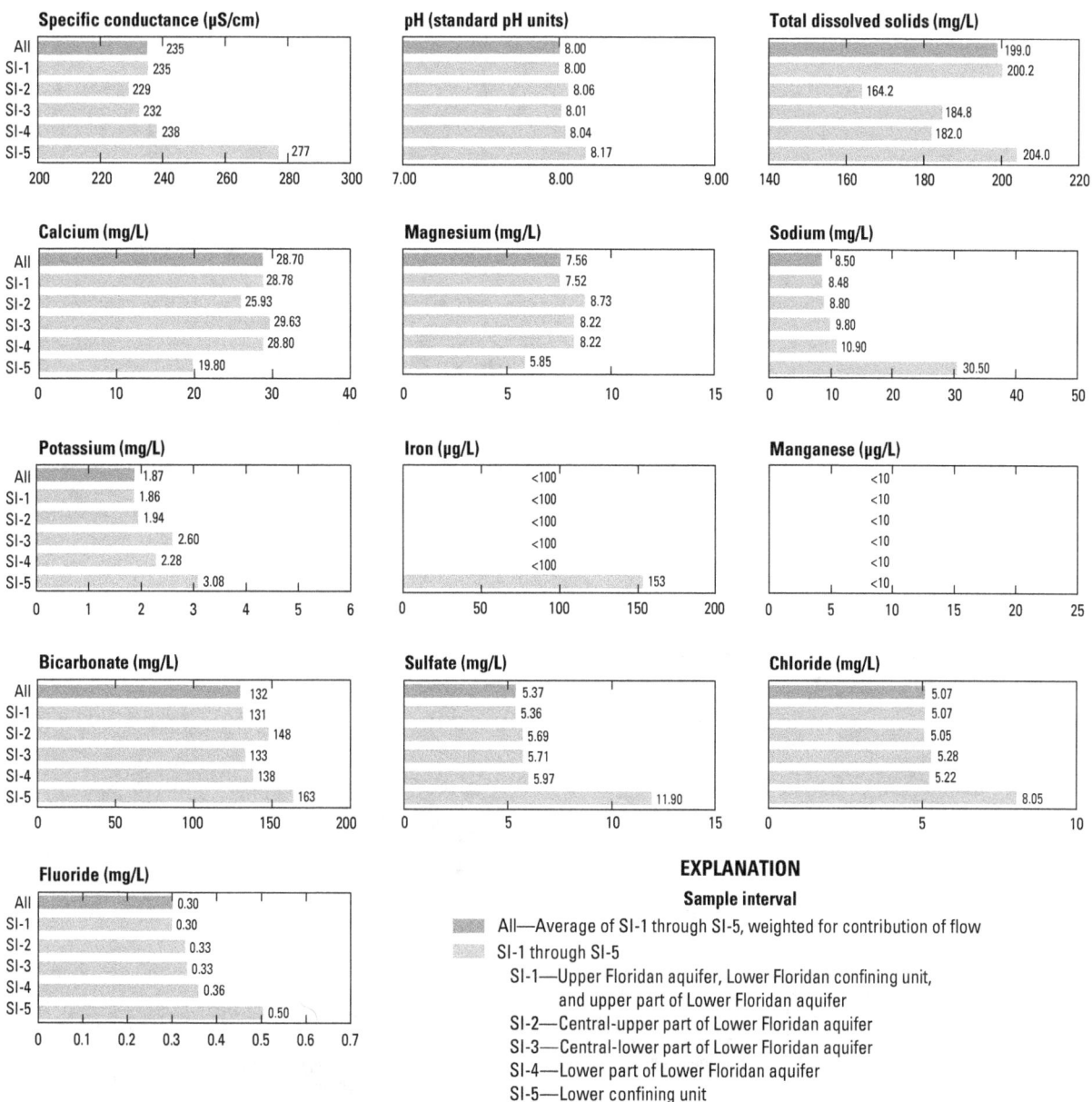

Figure 10. Water-quality values by sample interval. Samples were collected December 8, 2011, immediately following a flowmeter survey of the Floridan aquifer system at test hole 35Q069, Pooler, Georgia. Composite water-sample lab-analysis results and flowmeter-survey data (contribution of water-bearing zones) were used to calculate water quality from sample intervals. Sample-interval locations are shown in figure 7. Sample-interval specifications are listed in table 5. [μS/cm, microsiemens per centimeter at 25 degrees Celsius; mg/L, milligrams per liter; μg/L, micrograms per liter; SI, sample interval]

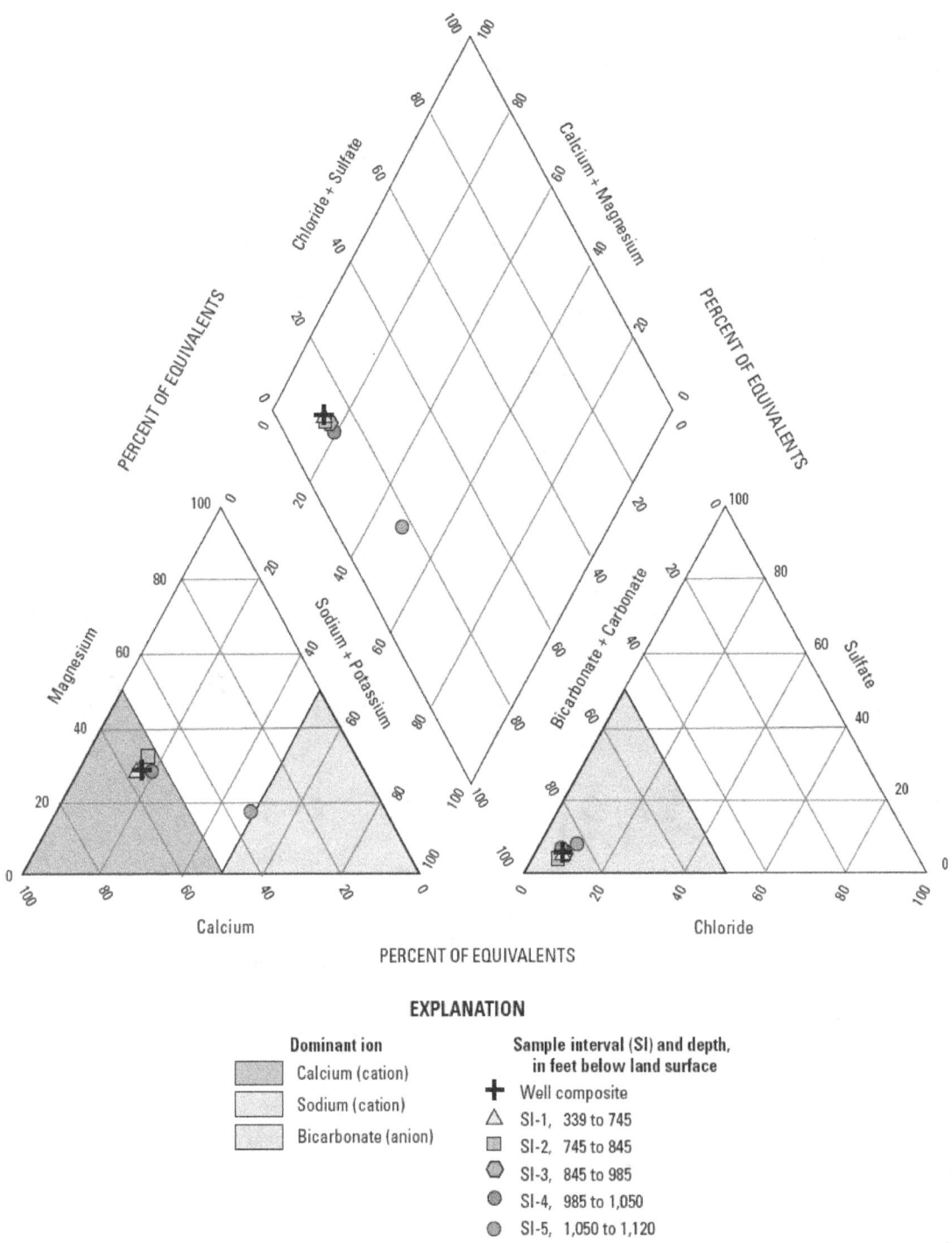

Figure 11. Piper diagram of water quality from sample intervals from test hole 35Q069, Pooler, Georgia, December 8, 2011. Composite water-sample lab-analysis results and flowmeter-survey data (contribution of water-bearing zones) were used to calculate water quality from sample intervals.

Hydrogeology and Water Quality

Hydrogeologic units of the Floridan aquifer system were distinguished by differences in flow contribution, lithology, geophysical characteristics, and water quality. Miller (1986) provided a regional definition of the Floridan aquifer system on the basis of widely spaced stratigraphic and borehole geophysical data in the coastal area of Georgia and South Carolina. Newly collected hydrogeologic and water-quality data were used to help refine Miller's (1986) regional definition of the Floridan aquifer system (Williams and Gill, 2010). The following sections describe the depths and hydraulic characteristics of hydrogeologic units that form the Floridan aquifer system at the Pooler test site.

Drillers and geophysical logs indicate a sediment sequence within the top 262 ft of test hole 35Q069 consisting of clay, sand, and some gravel. This clastic sequence represents a combination of the surficial and Brunswick aquifer systems and is part of the confining unit above the Floridan aquifer system (fig. 3). Varying amounts of phosphate were identified in the drill cuttings in the upper 262 ft of sediment. Below the 262 ft of clastic deposits is a 39-ft transition zone to limestone, consisting of a mix of shells, gravel, and sand. The limestone extends from 301 to 1,040 ft in depth and is typically moderately indurated and fossiliferous. Zones of poor induration exist at depths of 333–370, 430–470, 505–515, 560–570, 705–750, 900–910, and 960–980 ft. Limestone was found to contain some gypsum deposits at depths of 630–640, 780–790, and 820–830 ft. Drill cuttings collected from depths between 1,040 and 1,090 ft exhibit a mixed lithology of poorly indurated limestone and brown silt. This deeper unit is associated with increased natural-gamma radiation and decreased resistivity that extends to near the bottom of the hole at 1,120 ft. Williams and Gill (2010) interpreted these geophysical-log signatures in a test hole in Effingham County, Ga., to be the lower confining unit, not included as part of the Floridan aquifer system. Drill cuttings from depths between 1,090 and 1,120 ft were not inspected.

Based on the flowmeter survey, driller cuttings, and geophysical logs, constituent limestone units of the Floridan aquifer system at the Pooler test site extend in depth from 333 to 1,040 ft at the following depth intervals: Upper Floridan confining unit, 301–333 ft; UFA, 333–515 ft; LFCU, 515–702 ft; and LFA, 702–1,040 ft.

Upper Floridan Aquifer

The UFA at well 35Q069 mostly consists of lower Oligocene and upper Eocene carbonate units that include high permeability zones (fig. 3). The top of the UFA (at a depth of 333 ft) consists of the lower Oligocene Suwannee Limestone and corresponds to the top of loose, shelly limestone. An impermeable limestone unit of the lower Oligocene Lazaretto Creek Formation overlies the top of the aquifer and functions as the Upper Floridan confining unit. The base of the Tiger Leap Formation serves as the top of Upper Floridan confining unit, which lies just below the top of the Oligocene deposits. The top of the Oligocene corresponds to a spike in the natural-gamma log called the "C-marker" (Wait, 1965; Gregg and Zimmerman, 1974; Clarke and others, 1990). The thickest part of the aquifer consists of the upper Eocene Ocala Limestone, which is characterized by low natural-gamma radiation, the top of which is called the "D-marker" (Wait, 1965; Gregg and Zimmerman, 1974; Clarke and others, 1990). The base of the UFA (at a depth of 515 ft) is close to the contact between the upper Eocene Ocala Limestone and the middle Eocene Avon Park Formation, and corresponds to the base of a thin, porous limestone and the main water-bearing unit.

Regional maps showing the depth and thickness of the geologic units that form the Floridan aquifer system (Williams and Gill, 2010) indicate that the UFA underlying the Pooler test site has a depth interval of 370–540 ft below land surface. Lithologic, geophysical, and flowmeter data collected from well 35Q069 (fig. 3) were used to refine this depth interval to 333–515 ft.

Flowmeter Survey

On December 8, 2011, a borehole flowmeter survey was completed in the test hole for well 35Q069 in the depth interval between 339 and 1,110 ft below land surface at the Pooler test site while pumping at a rate of 708 gal/min. The survey indicated that 92.4 percent (654.2 gal/min) of the total flow originated from the UFA, and the remaining 7.6 percent (53.8 gal/min) was derived from the underlying LFCU and the LFA (fig. 6). Of the 654.2 gal/min contributed by the UFA, 6.3 percent of the flow was produced in the 339–350 ft interval, and 86.1 percent of the flow was produced in the 375–515 ft interval. The dominance of water-bearing potential in the UFA also was found at Hunter Army Airfield (Clarke and others, 2010) and Fort Stewart (table 8; Clarke and others, 2011; Gonthier, 2011). At all three sites, the majority of the water-bearing potential is located below the "D-marker" within the Ocala Limestone. Unlike test holes at Hunter Army Airfield and Fort Stewart, the test hole for well 35Q069 at the Pooler test site was cased above the "D-marker" and contained a relatively minor water-bearing zone just above the "D-marker." Relatively thin, distinct water-bearing zones in the UFA (below the "D-marker") were observed at Hunter Army Airfield. In contrast, more diffuse zones were observed at Fort Stewart and Pooler. One subtle difference between the main water-bearing zones in the UFA at Fort Stewart and Pooler is that water-bearing potential decreases with depth at Fort Stewart (Gonthier, 2011, p. 6) whereas water-bearing potential does not appear to change with depth at Pooler.

Table 8. Flowmeter survey results in percent contribution of flow from the Upper Floridan aquifer, Lower Floridan confining unit, and Lower Floridan aquifer from selected studies in the vicinity of Pooler, Georgia, 2009–2012.

[—, no measurement]

Hydrogeologic unit	Percent contribution of flow		
	Fort Stewart[a]	Pooler (this study)	Hunter Army Airfield[b]
Upper Floridan aquifer	92.3	92.4	83.6
Lower Floridan confining unit	3.1	3.0	4.3
Lower Floridan aquifer	4.6	4.6	11.7
Lower confining	—	0	0.6

[a]Gonthier, 2011.

[b]Clarke and others, 2010.

Hydraulic Properties

The transmissivity and storage coefficient of the UFA were estimated by analyzing water-level data collected during the 24-hour aquifer test performed during March 27–28, 2012, by pumping UFA well 35Q070 at a rate of 285 gal/min; and from the 72-hour aquifer test performed during April 16–19, 2012, by pumping LFA well 35Q069 at a rate of about 783 gal/min. Details of the aquifer tests, analysis, and results are discussed in appendix 2, and aquifer-test results are listed in table 4. Drawdown within pumped UFA well 35Q070 during the 24-hour aquifer test exhibited turbulence and an instantaneous water-level decline. The typical, gradual increases in drawdown with time that were used to determine hydraulic properties of the aquifer were small compared to nonaquifer-test influences that remained after filtering, making the drawdown data difficult to analyze with accuracy. A transmissivity estimate for the UFA of 30,000 ft²/d (fig. 2–9 in appendix 2, table 4) was computed using the Cooper-Jacob straight-line method (Cooper and Jacob, 1946).

Transmissivity and storage coefficient were also estimated by simulating the response of both wells to the 72-hour aquifer test at pumped LFA well 35Q069 (MODFLOW-96 and MODOPTIM as discussed in appendix 2). Estimated transmissivity and storage coefficient of the UFA at observation well 35Q070 were 46,000 ft²/d and 5.9×10^{-4}, respectively.

The accepted transmissivity for the UFA is the value that was estimated from the simulation of the response to the 72-hour aquifer test of LFA well 35Q069 (46,000 ft²/d). The drawdown data for UFA observation well 35Q070 in response to the 72-hour aquifer test contains much less noise than the drawdown data in response to the 24-hour aquifer test. Therefore, the results of the simulation of the 72-hour aquifer test are more reliable and less error prone than those of the 24-hour aquifer test. Also, the estimated transmissivity from the slug tests open to the entire Floridan aquifer system (58,000 ft²/d, see appendix 1) is close to the estimated transmissivity of the

entire Floridan aquifer system (51,000 ft²/d), based on the simulation of the 72-hour aquifer test. Considering a model thickness of 183 ft, the K_h and specific storage of the UFA totaled 250 ft/d and 3.2×10^{-6} ft^{-1}, respectively.

Water Quality

Water quality in the UFA is determined from sample interval one (SI-1), which is the interval between two composite grab samples taken at a depth of 339 and 745 ft and the specific conductance of drilling fluid. SI-1 spans all of the UFA and LFCU and a small part of the LFA thickness. Greater than 96 percent of the flow in this interval originates from the UFA (table 5). As a result, it is relatively safe to assume that two water-bearing zones below the base of the UFA do not influence the water chemistry of SI-1. Furthermore, the three sample intervals directly below SI-1 (SIs 2–4) have water quality similar to that of SI-1 (fig. 10, table 6). Values of specific conductance of drilling fluid fluctuated about 240 microsiemens per centimeter at 25 degrees Celsius (µS/cm) from the top of data collection at the 410 ft depth to the central part of the LFA at about the 890 ft depth (fig. 3).

Water type for SI-1 and the total composite flow is calcium bicarbonate (fig. 11). Specific conductance in water from SI-1 was 235 µS/cm, similar to that of the previously sampled drilling fluid and that of SIs 2-4. Major ions were present in low concentrations in water from SI-1, and the concentration of total dissolved solids was 200.2 milligrams per liter (mg/L). Concentrations of major ions, iron, and manganese within the UFA were well below water-quality criteria established by the U.S. Environmental Protection Agency for drinking water and health advisories (table 6; U.S. Environmental Protection Agency, 2011).

Lower Floridan Confining Unit

The LFCU at well 35Q069 consists of chalky and glauconitic limestone in the uppermost part of the middle Eocene Avon Park Formation between depths of 515 and 702 ft (figs. 2 and 3). Thickness and hydraulic conductivity of the confining unit control the rate of interaquifer leakage between the UFA and LFA.

The LFCU is 187 ft thick at the Pooler test site compared to 207 ft thick at Fort Stewart (Clarke and others, 2011) located 23 mi to the southwest, and 143 ft thick at Hunter Army Airfield (Clarke and others, 2010) located 7 mi to the east-southeast. The LFCU contains carbonate sediments of slightly lower permeability than those of the LFA. This unit is similar in lithology to overlying and underlying rock units, which precluded identification of the confining unit during drilling. Following construction of the test hole for well 35Q069, the thickness and depths of the confining unit were assessed by using borehole geophysical logs, the results of a flowmeter survey, and the deepest packer-isolated slug test.

Flowmeter Survey

The flowmeter survey in the test hole for well 35Q069 (fig. 6) indicated that the LFCU contributed little water to the overall flow in the test hole. In particular, continuous vertical sections of limestone between depths of 520 and 550 ft and between 620 and 702 ft contributed no detectable amounts of water during test pumping at 708 gal/min. Within the confining unit, a single water-bearing zone between depths of 550 and 620 ft yielded 21.3 gal/min, or 3.0 percent of the total test-hole flow, during the flowmeter survey.

Hydraulic Properties

The vertical hydraulic conductivity (K_v) and porosity of the LFCU were determined by hydraulic analysis of the core at three intervals: 564.3–569.8, 600.0–605.0, and 640.0–648.4 ft (table 2). Values of K_v ranged from 0.57 to 1.67 ft/d. These values compare with the lower end of the full range of reported hydraulic-conductivity values for karst limestone (Freeze and Cherry, 1979; 0.21 to 3,850 ft/d). Porosity values for all three cores were about 0.33 and are within the reported ranges for limestone and sandy clay (Heath, 1983). The K_v of the cores may not fully represent the K_v of the confining unit because of the small volumes that the cores represent.

The K_h of the LFCU was determined by completing packer-isolated slug tests at three separate intervals— 612–618, 642–648, and 690–696 ft (fig. 3)—prior to installation of the 8-inch-diameter casing to complete the test hole, as LFA well 35Q069. The K_h values for the three intervals ranged from 0.5 to 10 ft/d (table 3). The packer-isolated interval 690–696 ft had a K_h of 0.5 ft/d. Somewhere between the low hydraulic conductivity at 696 ft and the top of a water-bearing unit at 705 ft is the base of the LFCU. The base of the LFCU in the Pooler test site is determined to be at a depth of 702 ft, about midway between 696 and 705 ft.

The transmissivity and storage coefficient of the LFCU was estimated by analyzing water-level data collected during the 72-hour aquifer test performed on April 16–19, 2012, by pumping LFA well 35Q069 at a rate of about 783 gal/min. The response of both UFA well 35Q070 and LFA well 35Q069 to the aquifer test was simulated using MODFLOW-96 and MODOPTIM (as discussed in appendix 2). A transmissivity value and a storage coefficient for the LFCU of 700 ft²/d and 4.5×10^{-4}, respectively, were used to calibrate the model. Considering a model thickness of 182 ft, the K_h and specific storage of the LFCU were 4 ft/d and 2.5×10^{-6} ft^{-1}, respectively. The value of K_h estimated from the simulation is in the middle range of values determined from the packer-isolated slug tests (table 3).

Water Quality

The UFA, LFCU, and the upper part of the LFA are located between two water-sample-collection depths. As a result, all flow contributed from the LFCU is mixed with the UFA and the upper part of the LFA to create water chemistry of SI-1. Because of the mixture of water in SI-1, determining the quality of water that is derived from the LFCU is not possible. Based on the similarity of water chemistry of the top four sample intervals (fig. 10), however, the quality of water derived from the LFCU (depth interval 550–620 ft) is most likely similar to that of the UFA and LFA. Values of specific conductance of drilling fluid within the LFCU fluctuated about 240 µS/cm (fig. 3).

Lower Floridan Aquifer

The LFA at well 35Q069 consists of chalky and glauconitic limestone in the upper part of the middle Eocene Avon Park Formation that is similar in lithology to overlying units (fig. 2). At well 35Q069, the top of the LFA is at a depth of 702 ft and extends to a depth of 1,040 ft. Williams and Gill (2010) described the hydrogeology of LFA test well 36Q330, located at Berwick Plantation in Chatham County 3.73 mi southeast of the Pooler test site. Geophysical logs indicate that the LFA at the Berwick Plantation well (36Q330) is between depths of 712 and 1,080 ft. The LFA is 338 ft thick at the Pooler test site, which is similar to thicknesses measured at Fort Stewart (388 ft), Hunter Army Airfield (377 ft), and Berwick Plantation (368 ft).

Flowmeter Survey

Water-bearing zones were identified in the LFA using two flowmeter surveys. Results of flowmeter testing of the Floridan aquifer system in the test hole for well 35Q069 performed on December 8, 2011, identified four water-bearing depth intervals in the LFA: 702–745, 745–800, 800–925, 984–1,015 ft (fig. 6). The LFA contributed about 32.5 gal/min, or 4.6 percent of the total flow.

A second flowmeter survey was performed in well 35Q069 on April 18, 2012, during the 72-hour aquifer test and extended from the base of the casing at a depth of 705 ft to near the bottom of the well at a depth of 1,110 ft. This survey identified five water-bearing depth intervals within the LFA: 705–745, 745–925, 925–984, 984–1,015, and 1,015–1,040 ft. The main water-bearing zone was identified at a depth interval of 745–925 ft, yielding 84.0 percent of the total flow. Water-bearing zones generally were diffuse. No flow was detected below a depth of 1,040 ft where the lower confining unit to the Floridan aquifer system is located.

Hydraulic Properties

The transmissivity of the LFA was estimated by analyzing water-level data collected during the 72-hour aquifer test performed April 16–19, 2012, by pumping LFA well 35Q069 at a rate of about 783 gal/min. The response of both UFA well 35Q070 and LFA well 35Q069 to the aquifer test was simulated using MODFLOW-96 and MODOPTIM (as discussed in appendix 2). A transmissivity of 4,000 ft²/d was used for the LFA to calibrate the model. Considering a model thickness of 338 ft, the K_h of the LFA was estimated to be 12 ft/d.

Water Quality

The quality of water in the LFA was initially evaluated during drilling of the test hole for well 35Q069 by measuring specific conductance of air-rotary drilling fluid from the 410–1,120-ft-depth interval (fig. 3). Values of specific conductance of drilling fluid between depths of 900 and 1,080 ft were slightly lower (fluctuating about 225 µS/cm) compared to those between depths of 410 and 890 ft (fluctuating about 240 µS/cm). Between depths of 1,080 and 1,120 ft, values of specific conductance of drilling fluid slightly increased, fluctuating about 230 µS/cm.

More detailed water quality of the LFA was determined from four composite grab samples taken at depths of 745, 845, 985, and 1,050 ft after the flowmeter survey, which was completed in the test hole for well 35Q069 on December 8, 2011. Using the water-quality results of these composite grab samples and mixing equation 2, water-quality properties were evaluated for flow contributed to the sample intervals between the depths from which composite grab samples were collected. Results of SI-2, SI-3, and SI-4 (table 6) encompass 88 percent of the flow contributed to the LFA and represent the water quality of the LFA. The remaining 12 percent of the flow contributed to the LFA went into SI-1. SI-5 provides some insight on the water quality of the lower confining unit, located directly below the Floridan aquifer system.

Water chemistry for SI-2, SI-3, and SI-4 are nearly identical (figs. 10 and 11). The three sample intervals had nearly the same concentrations of fluoride (0.33–0.36 mg/L) and pH (8.01–8.06). For the three sample intervals, concentrations ranged as follows: calcium 25.93–29.63 mg/L; magnesium 8.22–8.73 mg/L; chloride 5.05–5.28 mg/L; sulfate 5.69–5.97 mg/L; bicarbonate 133–148 mg/L; and specific conductance 229–238 µS/cm. The water type for SI-2, SI-3, and SI-4 is calcium bicarbonate (fig. 11).

Major ions were present in low concentrations in water from the three sample intervals. The maximum concentration of total dissolved solids measured 184.8 mg/L. Concentrations of major ions, iron, and manganese within the LFA were well below U.S. Environmental Protection Agency secondary drinking-water regulations or health-based criteria (table 6;

U.S. Environmental Protection Agency, 2011). Water-quality results of SIs 2–4 were similar to those of SI-1, indicating that water quality in the Floridan aquifer system is homogenous.

Results from SI-5 indicate that water chemistry in the lower confining unit is distinct from that of the Floridan aquifer system (figs. 10 and 11; table 6). Compared to sample intervals in the Floridan aquifer system, SI-5 contained higher concentrations of potassium, sodium, chloride, fluoride, sulfate, bicarbonate, and total dissolved solids; higher values of pH and specific conductance; and lower concentrations of calcium and magnesium. The 1,050-ft grab sample was the only sample to have a detectable concentration of iron (153 micrograms per liter [µg/L]), which was below the U.S. Environmental Protection Agency (2011) secondary drinking water regulation of 300 µg/L. The water type within SI-5 is sodium bicarbonate.

The chemical results of SI-5 may not fully represent the water quality in the lower confining unit because of the low contribution of flow during the flowmeter surveys. Some water from above may have mixed with water from the lower confining unit during collection of the grab sample. Despite these limitations, the chemical analyses indicate that water from the lower confining unit should not have a strong effect on the water quality in LFA well 35Q069. Despite higher concentrations of most major ions compared to those from the Floridan aquifer system, concentrations of major ions, iron, and manganese in water from SI-5 were below U.S. Environmental Protection Agency secondary drinking-water regulations or normal health-based criteria for (U.S. Environmental Protection Agency, 2011). The one minor exception was a sodium concentration of 30.5 mg/L, which exceeded the health-based value of 20 mg/L for individuals on a 500-milligram-per-day restricted sodium diet.

A composite water sample collected from completed LFA well 35Q069 near the end of the 72-hour aquifer test on April 19, 2012, was tested for a limited number of analytes (table 7). The total dissolved solids concentration measured 190 mg/L. Concentrations of analyzed constituents were similar though slightly higher than concentrations of analyzed constituents contained in water samples collected from SIs 1–4. Chloride concentrations in the water sample from the aquifer test measured 6.4 mg/L compared to about 5.1 mg/L in SIs 1–4. Fluoride and sulfate concentrations in the water sample from the 72-hour aquifer test were 0.68 and 14 mg/L, respectively. Maximum fluoride and sulfate concentrations in water samples collected immediately following the flowmeter survey on December 8, 2011, were 0.5 and 11.9 mg/L, respectively. Despite higher concentrations of some analytes, concentrations of major ions were well below U.S. Environmental Protection Agency secondary drinking-water regulations and health-based criteria (table 6; U.S. Environmental Protection Agency, 2011). Analytes that were not detected included iron, manganese, zinc, and nitrate plus nitrite.

Effect of Lower Floridan Aquifer Pumping on the Upper Floridan Aquifer

The effect of pumping LFA well 35Q069 on water levels in UFA well 35Q070 were evaluated by monitoring drawdown response during the 72-hour aquifer test performed during April 16–19, 2012. Observed water-level responses in the UFA and LFA wells as a result of pumping the LFA were determined by using water-level data that were filtered for tidal, barometric, and long-term trends and by following procedures described in appendix 2. In response to pumping LFA well 35Q069 for 72 hours at a rate of about 783 gal/min, drawdown in LFA well 35Q069 and UFA well 35Q070 reached 51.7 and 0.9 ft, respectively (figs. 2–5 and 2–6, appendix 2). This drawdown response created a hydraulic head gradient with the water level approximately 50 ft lower in the LFA than in the UFA. Such a hydraulic-head gradient leads to the potential for flow of water from the UFA into the LFA; however, the actual flow rate was not determined from aquifer-test data and is beyond the scope of this report.

Summary

To assess the water-supply potential of the Lower Floridan aquifer (LFA) at Pooler, Georgia, the U.S. Geological Survey (USGS) in cooperation with the City of Pooler, performed an investigation during 2011–2012 to determine the hydrogeology and water quality of the Floridan aquifer system and the potential effect that pumping from the LFA would have on the Upper Floridan aquifer (UFA). The study included construction of a test well in the UFA and a test well in the LFA, detailed site investigations, and hydraulic and water-quality characterization of the Floridan aquifer system. Based on the flowmeter survey, driller cuttings, and geophysical logs, hydrogeologic units of the Floridan aquifer system are present at the following depth intervals: Upper Floridan confining unit, 301–333; UFA 333–515; Lower Floridan confining unit (LFCU), 515–702; and LFA, 702–1,040 feet (ft).

The UFA at well 35Q069 mostly consists of lower Oligocene and upper Eocene carbonate units that include high permeability zones. A borehole flowmeter survey completed in the test hole for well 35Q069 open to the entire Floridan aquifer system indicated that 92.4 percent of the total flow of 708 gallons per minute (gal/min) originated from the UFA, and the remaining 7.6 percent was derived from the underlying LFCU and LFA. Most of the flow (86.1 percent) in the UFA was produced in the 375–515 ft interval. During the spring of 2012, 24- and 72-hour aquifer tests were performed to determine hydraulic properties of the UFA. Estimates of transmissivity were derived using the Cooper-Jacob method for the 24-hour aquifer test and simulated with MODOPTIM with MODFLOW for the 72-hour aquifer test. Transmissivity ranged from 30,000 to 46,000 feet squared per day (ft^2/d) for the 24-hour and 72-hour aquifer tests, respectively, with the higher value considered to be more representative of the actual transmissivity. Considering a model thickness of 187 ft, the horizontal hydraulic conductivity (K_h) and specific storage of the UFA were 250 feet per day (ft/d) and 3.2×10^{-6} per foot (ft^{-1}), respectively.

UFA water quality was determined from two borehole grab samples collected immediately following the flowmeter survey. The UFA has a calcium bicarbonate water type with low concentrations of major ions and a total dissolved solids concentration of 200.2 milligrams per liter (mg/L). Concentrations of major ions, iron, and manganese within the UFA were below U.S. Environmental Protection Agency secondary drinking-water regulations and health-based criteria.

The LFCU at well 35Q069 consists of chalky and glauconitic limestone in the uppermost part of the middle Eocene Avon Park Formation between depths of 515 and 702 ft. Thickness and hydraulic conductivity of the confining unit control the rate of leakage between the UFA and LFA. A flowmeter survey indicated that the LFCU contributed little to the overall flow in the test hole, with a single water-bearing zone at 550–620 ft yielding 21.3 gal/min, or 3.0 percent of the total test-hole flow. Values of vertical hydraulic conductivity (K_v) and porosity of the LFCU were determined by analyzing the core for hydraulic analysis at three intervals. Values of K_v from the sample ranged from 0.57 to 1.67 ft/d; porosity values for the three cores were about 0.33. A sample of values of horizontal hydraulic conductivity (K_h) of the LFCU was determined by completing packer-isolated slug tests at three separate intervals. The K_h values for the three intervals ranged from 0.5 to 10 ft/d.

The transmissivity and storage coefficient of the LFCU was estimated from the 72-hour aquifer test by pumping LFA well 35Q069 at a rate of about 783 gal/min. A transmissivity and storage coefficient for the LFCU of 700 ft^2/d and 4.5×10^{-4}, respectively, were used to calibrate the model. Considering a model thickness of 182 ft, the K_h and specific storage of the LFCU was 4 ft/d, and 2.5×10^{-6} ft^{-1}, respectively. The value of K_h estimated from the model simulation is in the middle of the range of values determined from the packer-isolated slug tests.

The LFA at well 35Q069 consists of chalky and glauconitic limestone in the upper part of the middle Eocene Avon Park Formation that is similar in lithology to overlying units. Results of flowmeter testing identified five water-bearing depth intervals within the LFA: 705–745, 745–925, 925–984, 984–1,015, and 1,015–1,040 ft, with the interval 745–925 ft contributing 84.0 percent of the total flow of about 783 gal/min. Water-bearing zones generally were diffuse. No flow was detected below a depth of 1,040 ft where the lower confining unit is located. Transmissivity of the LFA was estimated through model simulation using data from a 72-hour aquifer test where LFA well 35Q069 was pumped at a rate of about 783 gal/min. Estimated transmissivity for the LFA was 4,000 ft^2/d. Considering a model thickness of 338 ft, the K_h of the LFA was estimated at 12 ft/d.

The quality of water in the LFA was determined from four grab samples collected during flowmeter testing of the test hole and from a composite sample collected from the completed LFA well near the end of the 72-hour aquifer test. Water from the LFA is a calcium bicarbonate type, with low concentrations of major ions, and a total dissolved solids concentration of 190 mg/L. Concentrations of all constituents were below U.S. Environmental Protection Agency secondary drinking-water regulations and health-based criteria.

The effect of pumping the LFA well 35Q069 on water levels in UFA well 35Q070 were evaluated by monitoring drawdown response during the 72-hour aquifer test. Observed water-level responses in the UFA and LFA wells as a result of pumping the LFA were determined by using water-level data that were filtered for tidal, barometric, and long-term trends. Drawdown in LFA pumped well 35Q069 was 51.7 ft, whereas drawdown in UFA well 36Q070 was 0.9 ft after 72 hours of pumping at a rate of about 783 gal/min.

Selected References

American Society for Testing and Materials, 2008, ASTM D5079-08 Standard practices for preserving and transporting rock core samples: Book of Standards, Volume 04.08, 7 p.

American Society for Testing and Materials, 2010, ASTM D5084-10 Standard test methods for measurement of hydraulic conductivity and saturated porous materials using a flexible wall permeameter: Book of Standards, Volume 04.08, 23 p.

Batu, Vedat, 1998, Aquifer hydraulics—A comprehensive guide to hydrogeologic data analysis: New York, N.Y., John Wiley & Sons, Inc., 727 p.

Bouwer, Herman, and Rice, R.C., 1976, A slug test for determining hydraulic conductivity of unconfined aquifers with completely or partially penetrating wells: Water Resources Research, v. 12, no. 3, p. 423–428.

Clark, W.Z., and Zisa, A.C., 1976, Physiographic map of Gerogia: Atlanta, Georgia Geological Survey, 1 sheet, scale 1:2,000,000.

Clarke, J.S., 2003, The surficial and Brunswick aquifer systems—Alternative ground-water resources for coastal Georgia, in Hatcher, K.J., ed., Proceedings of the 2003 Georgia Water Resources conference, April 23–24, 2003: Athens, University of Georgia, CD–ROM.

Clarke, J.S., Cherry, G.C., Gonthier, J.G., 2011, Hydrogeology and water quality of the Floridan aquifer system and effects of Lower Floridan aquifer pumping on the Upper Floridan aquifer at Fort Stewart, Georgia: U.S. Geological Survey Scientific Investigations Report 2011–5065, 59 p., available at http://pubs.usgs.gov/sir/2011/5065/.

Clarke, J.S., Hacke, C.M., and Peck, M.F., 1990, Geology and ground-water resources of the coastal area of Georgia: Atlanta, Georgia Geologic Survey Bulletin 113, 106 p.

Clarke, J.S., and Krause, R.E., 2000, Design, revision, and application of ground-water flow models for simulation of selected water-management scenarios in the coastal area of Georgia and adjacent parts of South Carolina and Florida: U.S. Geological Survey Water-Resources Investigations Report 00–4084, 93 p.

Clarke, J.S., Leeth, D.C., Taylor-Harris, Da'Vette, Painter, J.A., and Labowski, J.L, 2004, Summary of hydraulic properties of the Floridan aquifer system in coastal Georgia and adjacent parts of South Carolina and Florida: U.S. Geological Survey Scientific Investigations Report 2004–5264, 52 p.

Clarke, J.S., Williams, L.J., and Cherry, G.C., 2010, Hydrogeology and water quality of the Floridan aquifer system and effect of Lower Floridan aquifer pumping on the Upper Floridan aquifer at Hunter Army Airfield, Chatham County, Georgia: U.S. Geological Survey Scientific Investigations Report 2010–5080, 56 p., available at http://pubs.usgs.gov/sir/2010/5080/.

Clemo, T., 2002, MODFLOW-2000 for cylindrical geometry with internal flow observations and improved water table simulation: Technical Report BSU CGISS 02-01, Boise State University, Boise, Idaho, 29 p.

Cooper, H.H., and Jacob, C.E., 1946, A generalized graphical method for evaluating formation constants and summarizing well field history: American Geophysical Union Transactions, v. 27, p. 526–534.

Counts, H.B., and Donsky, Ellis, 1963, Salt-water encroachment, geology, and ground-water resources of the Savannah area, Georgia and South Carolina: U.S. Geological Survey Water-Supply Paper 1611, 100 p.

Cunningham, W.L., and Schalk, C.W., compilers, 2011, Groundwater technical procedures of the U.S. Geological Survey: U.S. Geological Survey Techniques and Methods 1—A1, 151 p., available at http://pubs.usgs.gov/tm/1a1/.

Doherty, John, 2005, Manual for Version 10 of PEST: Brisbane, Australia, Watermark Numerical Computing.

Falls, W.F., Harrelson, L.G., Conlon, K.J., and Petkewich, M.D., 2005, Hydrogeology, water quality, and water-supply potential of the Lower Floridan aquifer, coastal Georgia, 1999–2002: U.S. Geological Survey Scientific Investigations Report 2005–5124, 89 p., 1 pl.

Fanning, J.L, and Trent, V.P., 2009, Water use in Georgia by county for 2005; and water-use trends, 1980–2005: U.S. Geological Survey Scientific Investigations Report 2005–5002, 186 p.

Freeze, R.A., and Cherry, J.A., 1979, Groundwater: Englewood Cliffs, N.J., Prentice Hall, 604 p.

Furlow, J.W., 1969, Stratigraphy and economic geology of the eastern Chatham County phosphate deposit: Georgia Department of Mines, Mining, and Geology Bulletin 82, 40 p.

Garber, M.S., and Koopman, F.C., 1968, Methods of measuring water levels in deep wells: U.S. Geological Survey Techniques of Water-Resources Investigations, book 8, chapter A1, 23 p.

Gonthier, G.J., 2011, Summary of hydrologic testing of the Floridan aquifer system at Fort Stewart, Georgia: U.S. Geological Survey Open-File Report 2011–1020, 28 p.

Gonthier, G.J., 2012, Hydrogeologic characteristics and water quality of a confined sand unit in the surficial aquifer system, Hunter Army Airfield, Chatham County, Georgia: U.S. Geological Survey Scientific Investigations Report 2012–5082, 14 p., available at http://pubs.usgs.gov/sir/2012/5082/.

Gregg, D.O., and Zimmerman, E.A., 1974, Geologic and hydrologic control of chloride contamination in aquifers at Brunswick, Glynn County, Georgia: U.S. Geological Water-Supply Paper 2029-D, 44 p.

Halford, K.J., 2006a, Documentation of a spreadsheet for time-series analysis and drawdown estimation: U.S. Geological Survey Scientific Investigations Report 2006–5024, 38 p.

Halford, K.J., 2006b, MODOPTIM: A general optimization program for ground-water flow model calibration and ground-water management with MODFLOW: U.S. Geological Survey Scientific Investigations Report 2006–5009, 62 p.

Halford, K.J., and Kuniansky, E.L., 2002, Documentation of spreadsheets for the analysis of aquifer-test and slug-test data: U.S. Geological Survey, Open-File Report 2002–197, 51 p.

Harbaugh, A.W., Banta, E.R., Hill, M.C., and McDonald, M.G., 2000, MODFLOW-2000, the U.S. Geological Survey modular ground-water flow model—User guide to modularization concepts and the ground-water flow process: U.S. Geological Survey Open-File Report 00–92, 121 p.

Harbaugh, A.W., and McDonald, M.G., 1996, Progammer's documentation for MODFLOW-96, an update to the U.S. Geological Survey modular finite difference ground-water flow model: U.S. Geological Survey Open-File Report 96–486, 220 p.

Heath, R.C., 1983, Basic ground-water hydrology: U.S. Geological Survey Water-Supply Paper 2220, 84 p.

Hill, M.C., 1998, Methods and guidelines for effective model calibration: U.S. Geological Survey Water-Resources Investigations Report 98–4005, 90 p.

Kendall, Carol, and Caldwell, E.A., 1998, Fundamentals of isotope geochemistry, in Kendall, Carol, and McDonnell, J.J., eds, 1998, Isotope tracers in catchment hydrology: Amsterdam, The Netherlands, Elsevier Science B.V., 839 p.

Langevin, C.D., 2008, Modeling axisymmetric flow and transport: Ground Water, v. 46, no. 4, p. 579–590.

McDonald, M.G., and Harbaugh, A.W., 1988, A modular three-dimensional finite-difference ground-water flow model: U.S. Geological Survey Techniques of Water-Resources Investigations, book 6, chapter A1, 576 p.

Miller, J.A., 1986, Hydrologic framework of the Floridan aquifer system in Florida, and in parts of Georgia, Alabama, and South Carolina: U.S. Geological Survey, Regional Aquifer-System Analysis, Professional Paper 1403-B, 91 p., 33 pls.

Payne, D.F., Abu Rumman, Malek, and Clarke, J.S., 2005, Simulation of ground-water flow in coastal Georgia and adjacent parts of South Carolina and Florida—Predevelopment, 1980 and 2000: U.S. Geological Survey Scientific Investigations Report 2005–5089, 91 p., available at http://pubs.usgs.gov/sir/2005/5089/.

Piper, A.M., 1944, A graphic procedure in the geochemical interpretation of water analyses: American Geophysical Union Transactions, v. 25, p. 914–923.

Poeter, E.P., and Hill, M.C., 1997, Inverse models: A necessary next step in ground-water modeling: Ground Water, v. 35, no. 2, p. 250–260.

Randolph, R.B., Pernik, Maribeth, and Garza, Reggina, 1991, Water-supply potential of the Floridan aquifer system in the coastal area of Georgia—A digital model approach: Atlanta, Georgia Geologic Survey Bulletin 116, 30 p.

Reily, T.E., and Harbaugh, A.W., 1993, Computer note: Simulation of cylindrical flow to a well using the U.S. Geological Survey modular finite-difference ground-water flow model: Ground Water, v. 31, no. 3, p. 489–494.

Rutledge, A.T., 1991, An axisymmetric finite-difference flow model to simulate drawdown in and around a pumped well: U.S. Geological Survey Water-Resources Investigations Report 90–4098, 33 p.

Stallman, R.W., 1971, Aquifer-test design observation, and data analysis: U.S. Geological Survey Techniques of Water-Resources Investigations, book 3, chap. B1, 26 p.

U.S. Census Bureau, 2002, Georgia: 2000 Census of population and housing, summary population and housing characteristics: U.S. Department of Commerce, PHC-1-12, variously paginated (about 428 p.).

U.S. Environmental Protection Agency, 2011, 2011 Edition of the drinking water standards and health advisories: U.S. Environmental Protection Agency, EPA 820-R-11-002, 18 p., available at http://water.epa.gov/action/advisories/drinking/upload/dwstandards2011.pdf.

van der Kamp, Garth, 1976, Determining aquifer transmissivity by means of well response tests—The underdamped case: Water Resources Research, v. 12, no. 1, p. 71–77.

Wait, R.L., 1965, Geology and occurrence of fresh and brackish ground water in Glynn County, Georgia: U.S. Geological Survey Water-Supply Paper 1613-E, 94 p.

Weems, R.E., and Edwards, L.E., 2001, Geology of Oligocene, Miocene, and younger deposits in the coastal area of Georgia: Atlanta, Georgia Geologic Survey Bulletin 131, 124 p.

Williams, L.J., 2010, Summary of hydrologic testing of the Floridan aquifer system at Hunter Army Airfield, Chatham County, Georgia: U.S. Geological Survey Open-File Report 2010–1066, 30 p., available at http://pubs.usgs.gov/of/2010/1066/.

Williams, L.J., and Gill, H.E., 2010, Revised hydrogeologic framework of the Floridan aquifer system in the northern coastal area of Georgia and adjacent parts of South Carolina: U.S. Geological Survey Scientific Investigations Report 2010–5158, 103 p.

Appendixes 1 and 2

Appendix 1. Slug tests, Pooler, Georgia, December 13–14, 2011

Slug tests were performed in the 10.3-inch-diameter test hole for well 35Q069 at three packer-isolated intervals within the Lower Floridan confining unit (LFCU) and within the entire test hole open to the Floridan aquifer system. Horizontal hydraulic conductivity (K_h) estimates of the LFCU within three packer-isolated intervals ranged from 0.5 foot per day (ft/d) in the deepest interval with a depth of 690–696 feet (ft; table 3), to 10 ft/d in the shallowest interval with a depth of 612–618 ft. Packer seals for each of the tests appeared tight, with leakage around the packer possibly affecting results of the falling-head slug test of the deepest interval. The estimated transmissivity of the Floridan aquifer system from the open hole slug tests totaled 58,000 feet squared per day (ft²/d).

Test Results at the 612- to 618-Foot Interval

Analysis of two slug tests were performed on the 612- to 618-ft packer-isolated interval on December 13, 2011, and resulted in comparable estimates of K_h of about 10 ft/d. For the first test, the initial static water level stood 46.2 ft below land surface. The slug was inserted into the water column at 3:29:26 p m. In response to slug insertion, the water level immediately spiked to about 6 ft above static water level followed by a drop to 3.5 ft above static water level within 5 seconds (fig. 1–1). Water-level recovery was within the resolution of the transducers (0.02 ft) about 4.5 minutes after the slug insertion. Packer seals appeared tight, as water levels above and below the packer-isolated interval did not exhibit a detectable response to the slug test. Within the packer-isolated interval, the smooth recovery typified the expected exponential decay of water-level displacement that was analyzed using the Bouwer and Rice (1976) method with a straight line on the analysis plot (fig. 1–2). The horizontal hydraulic conductivity (K_h) derived from these data was 9.6 ft/d.

The second slug test for the 612- to 618-ft packer-isolated interval was performed about 10 minutes after the slug was initially inserted by removing the slug from the water column at 3:39:32 p.m. (fig. 1–3). In response to slug removal, the water level dropped to 3.74 ft below static water level after about 8 seconds. Water-level recovery was within 0.02 ft a little more than 5 minutes after slug removal. Packer seals appeared tight, as water levels above and below the packer-isolated interval did not exhibit a detectable response to the slug test. Within the packer-isolated interval, smooth recovery typified the expected exponential decay of water-level displacement that was analyzed using the Bouwer and Rice (1976) method with a straight line on the analysis plot (fig. 1–4). The K_h derived from these data was 9.4 ft/d.

Test Results at the 642- to 648-Foot Interval

Analysis of two slug tests were performed on the 642- to 648-ft packer-isolated interval on December 14, 2011, and resulted in comparable estimates of K_h of about 3 ft/d. For the first test, the initial static water level stood 46.1 ft below land surface. The slug was inserted into the water column at 8:46:48 a.m. In response to slug insertion, the water level immediately spiked to 7.4 ft above static water level followed by a drop to 4.2 ft above static water level after 9 seconds (fig. 1–5). Water-level recovery was within 0.02 ft of static conditions about 13 minutes after slug insertion. Packer seals appeared tight, as water levels above and below the packer-isolated interval did not exhibit a detectable response to the slug test. Within the packer-isolated interval, the smooth recovery typified the expected exponential decay of water-level displacement that was analyzed using the Bouwer and Rice (1976) method with a straight line on the analysis plot (fig. 1–6). The K_h derived from these data was 3.4 ft/d.

The second slug test for the 642- to 648-ft interval was performed about 15 minutes after the slug was initially inserted by removing the slug from the water column at 9:01:08 a m. (fig. 1–7). In response to slug removal, the water level dropped to 4.52 ft below static water level after 9 seconds. Water-level recovery was within 0.02 ft of static water level about 11 minutes after slug removal. Packer seals appeared tight, as water levels above and below the packer-isolated interval did not exhibit a detectable response to the slug test. Within the packer-isolated interval, smooth recovery nearly typified the expected exponential decay of water-level displacement that was analyzed using the Bouwer and Rice (1976) method. Only a slight negative concavity in the water-level-recovery curve was exhibited on the analysis plot (fig. 1–8). The slight negative concavity in the curve is only apparent about 3.5 minutes after the start of the slug test and did not affect the estimate of K_h, which was determined to be 3.3 ft/d.

Test Results at the 690- to 696-Foot Interval

Analysis of a rising-head slug test were performed on the 690- to 696-ft packer-isolated interval on December 14, 2011, and resulted in an estimate of K_h of 0.5 ft/d. For the first test, the initial static water level stood 46.1 ft below land surface. The slug was inserted into the water column at 11:51:11 a m. In response to the slug insertion, the water level immediately spiked to 5.7 ft above static water level and then oscillated for about 22 seconds (fig. 1–9). Water-level recovery was within

0.02 ft about 18 minutes after slug insertion. The lower packer seal appeared to leak. Water levels above the packer-isolated interval did not exhibit a response to the slug test; however, in response to the slug insertion, water levels below the packer-isolated interval increased by about 0.02 ft above static water level for more than 5 minutes. This slight water-level increase indicates leakage between the packer-isolated interval and the lower transducer. Due to the response of the lower transducer, no hydraulic conductivity was estimated from this slug test. The air pressure to the packers was increased at the noon hour to prevent leakage around the lower packer during slug removal. The air-pressure increase caused a slight temporary water-level increase that occurred late in the slug-test recovery.

The second slug test for the 690- to 696-ft packer-isolated interval was performed about 18 minutes after the slug was initially inserted by removing the slug from the water column at 12:08:45 p m. (fig. 1–10). In response to slug removal, the water level dropped to 4.53 ft below static water level after 10 seconds. Water-level recovery was within 0.02 ft of static water level about 51 minutes after the slug removal. Packer seals appeared tight, as water levels above and below the packer-isolated interval did not exhibit a detectable response to the slug test. Within the packer-isolated interval, the recovery did not typify a perfect exponential decay of water-level displacement that was analyzed using the Bouwer and Rice (1976) method, but rather had a slightly negative concavity within the curve on the analysis plot (fig. 1–11). Consequently, the K_h was estimated by aligning analytical points to the recovery data during 5.5–15.7 minutes following slug removal. The estimated K_h for this slug test was 0.5 ft/d.

Test Results at the 339- to 1,130-Foot Interval

After slug tests were completed within packer-isolated intervals, the packers were deflated while the packer assembly remained at the 690- to 696-ft depth. The water column that was connected to the packer assembly was then connected to the Upper Floridan aquifer, LFCU, and Lower Floridan aquifer (the entire Floridan aquifer system) between depths of 339 and 1,130 ft. All three transducers were, therefore, hydraulically connected to the water column. Analysis of two slug tests performed on the entire Floridan aquifer system resulted in comparable estimates of transmissivity of about 58,000 ft²/d. For the first test, the initial static water level stood 46.0 ft below land surface. The slug was inserted into the water column on December 14, 2011, at 1:11:42 p m. Slug insertion caused identical water-level oscillations at the middle and lower transducers having an initial and maximum displacement of about 0.25 ft above static water level and a period of about 29 seconds.

The upper transducer measured a damped water-level response to slug insertion, attaining a maximum water-level displacement of about 0.10 ft, or about 40 percent to that of the middle and lower transducers. Furthermore, water levels at the upper transducer responded immediately to slug insertion with oscillations having the same period (29 seconds) as that of the water levels measured with the middle and lower transducers. Maximum water-level displacement, however, was achieved about 5 seconds after maximum water-level displacements were recorded at the middle and lower transducers.

The second slug test for the 339- to 1,130-ft interval was performed when the amplitude of the oscillation from the previous test damped to an undetectable amount (by 1:17:58 p m.), slightly more than 4 minutes after slug insertion. Removing the slug at this time resulted in a water-level response that virtually mirrored that of the slug insertion (fig. 1–12).

The two slug tests for the 339- to 1,130-ft open interval were analyzed using the van der Kamp (1976) method (figs. 1–13 and 1–14). A storage coefficient value for the Floridan aquifer system of 1.7×10^{-3} was used based on the 72-hour aquifer-test results (appendix 2). A borehole diameter of 10.2 inches was used based on the average borehole diameter for the open interval. The transmissivity estimates from the rising (insertion) and falling (removal) slug tests were 54,000 and 62,000 ft²/d, respectively.

The estimate of transmissivity is affected by parameter sensitivity and by likely attenuation of the slug influence. Transmissivity estimates are sensitive to both the storage coefficient and the borehole diameter that is used in the van der Kamp (1976) method. As an example, increasing the borehole diameter from 10.2 to 10.5 inches and decreasing the storage coefficient from 1.7×10^{-3} to 1.1×10^{-3} in the falling-head slug test increases the estimated transmissivity from 54,000 to 60,000 ft²/d. The lower maximum displacement response of the upper transducer, compared to the transducers that are closer to the slug influence near the center of the packer assembly, indicates that the slug influence does not evenly affect the entire borehole. As a result, the estimate of transmissivity does not necessarily reflect an even measurement of the entire Floridan aquifer system.

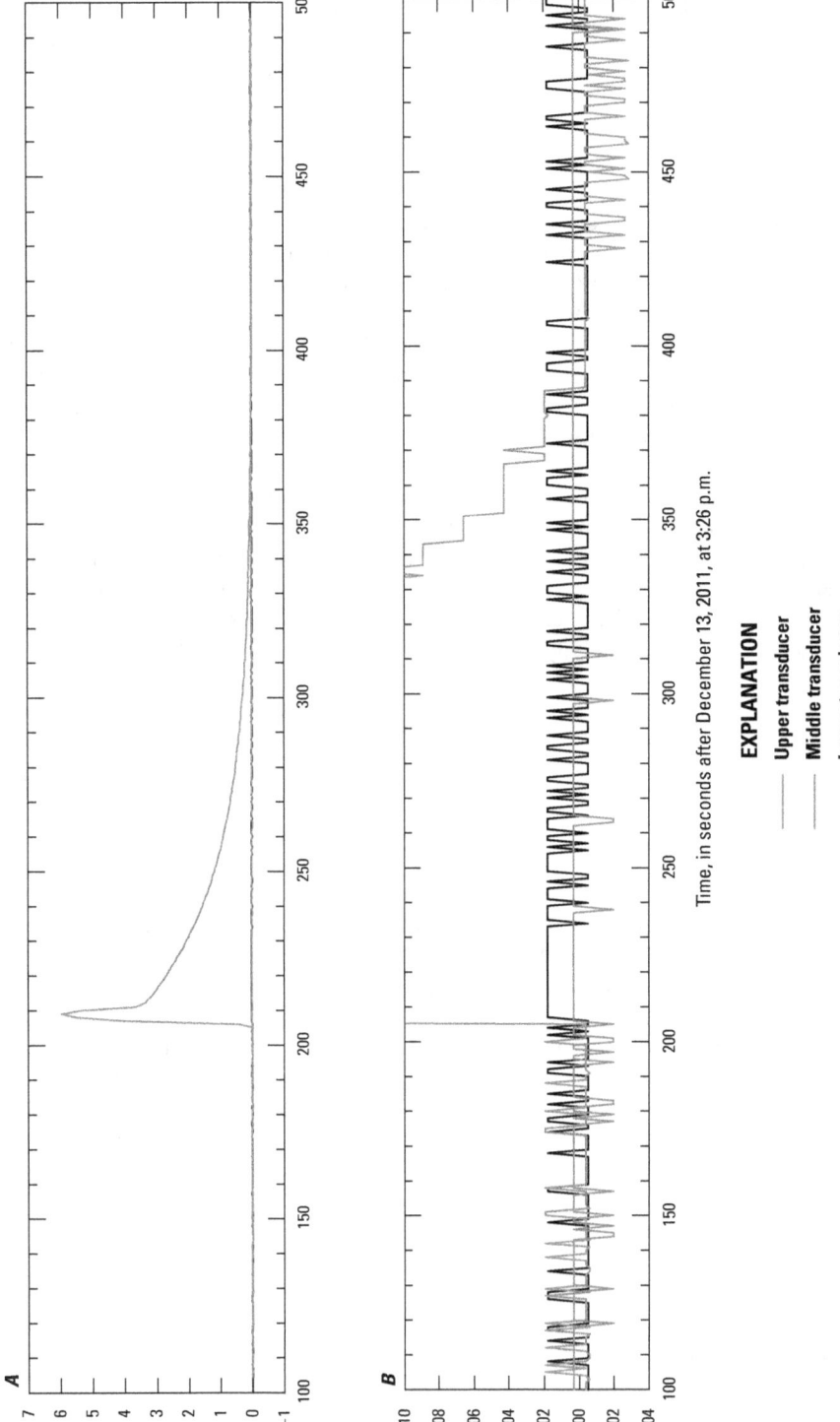

Figure 1–1. Water-level response to inserting slug into the water column connected to the packer-isolated interval at 612 to 618 feet below land surface, Lower Floridan confining unit, test hole 350069, Pooler, Georgia, December 13, 2011. Static water level was 46.2 feet below land surface. *A,* Complete vertical extent of water-level response. *B,* Larger vertical scale of water-level response.

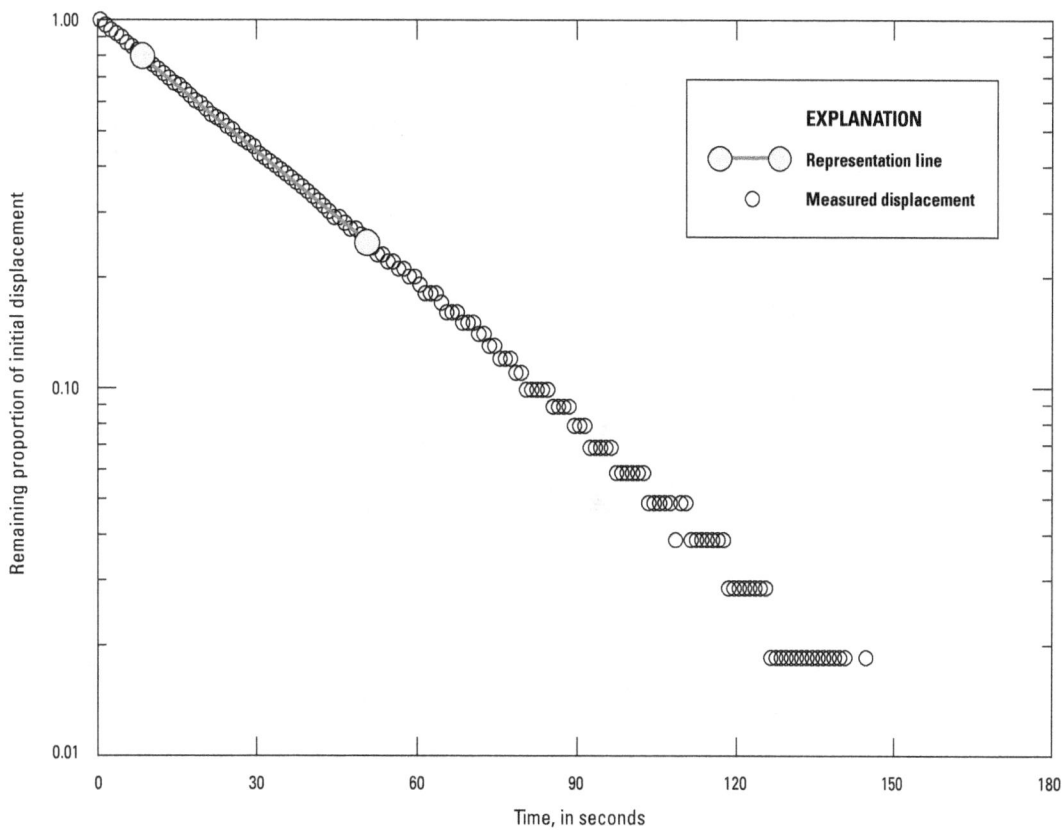

Figure 1–2.   Semi-log plot of water-level recovery from inserting slug into the water column connected to the packer-isolated interval at 612 to 618 feet below land surface, Lower Floridan confining unit, test hole 35Q069, Pooler, Georgia, December 13, 2011. Representation line was used to determine the horizontal hydraulic conductivity of the packer-isolated interval, using the Bouwer and Rice (1976) method. Modified from a graph on a spreadsheet from Halford and Kuniansky (2002).

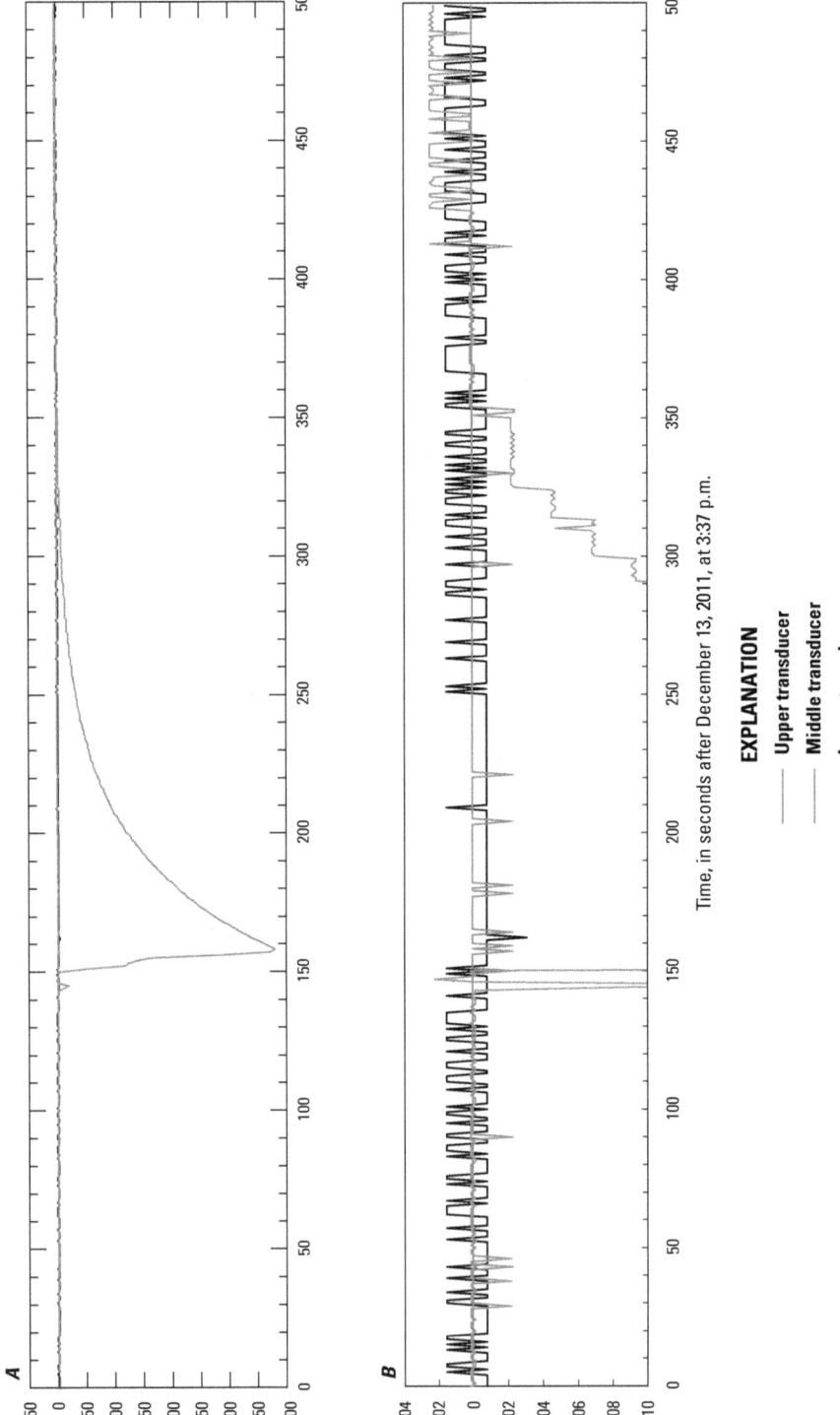

Figure 1–3. Water-level response to pulling a slug from the water column connected to the packer-isolated interval at 612 to 618 feet below land surface, Lower Floridan confining unit, test hole 35Q069, Pooler, Georgia, December 13, 2011. Static water level was 46.2 feet below land surface. *A*, Complete vertical extent of water-level response. *B*, Larger vertical scale of water-level response.

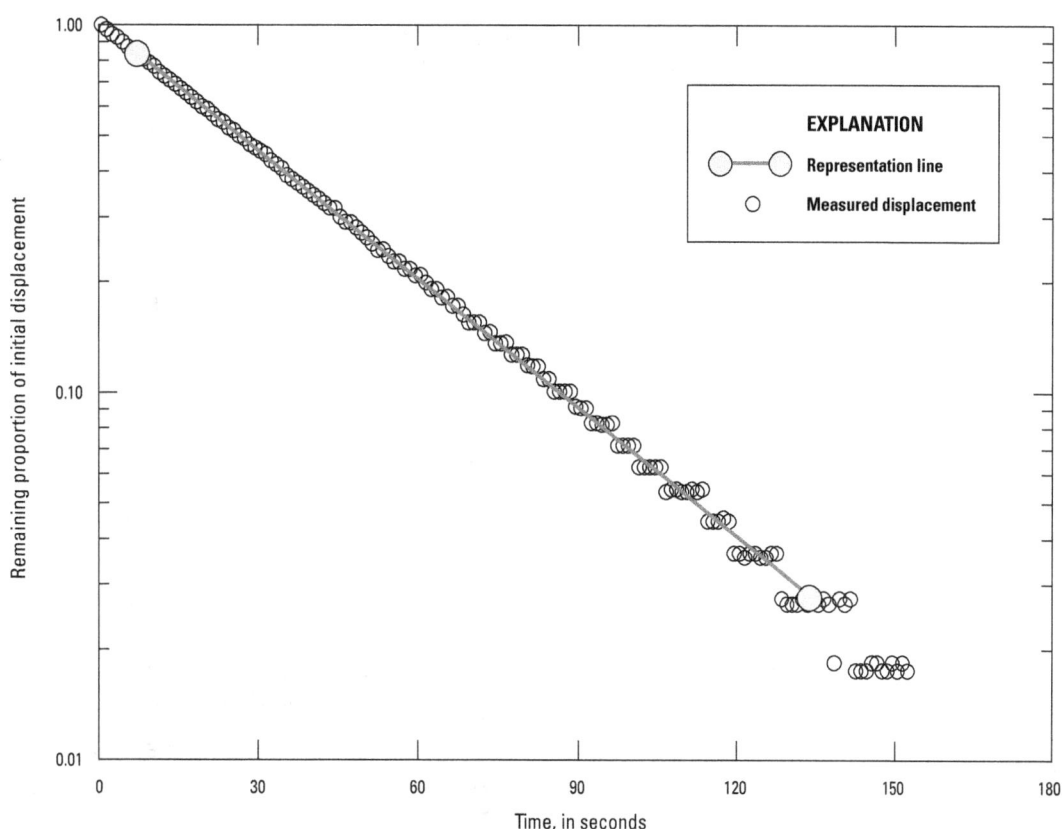

Figure 1–4.   Semi-log plot of water-level recovery from pulling slug from the water column connected to the packer-isolated interval at 612 to 618 feet below land surface, Lower Floridan confining unit, test hole 35Q069, Pooler, Georgia, December 13, 2011. Representation line was used to determine the horizontal hydraulic conductivity of the packer-isolated interval, using the Bouwer and Rice (1976) method. Modified from a graph on a spreadsheet from Halford and Kuniansky (2002).

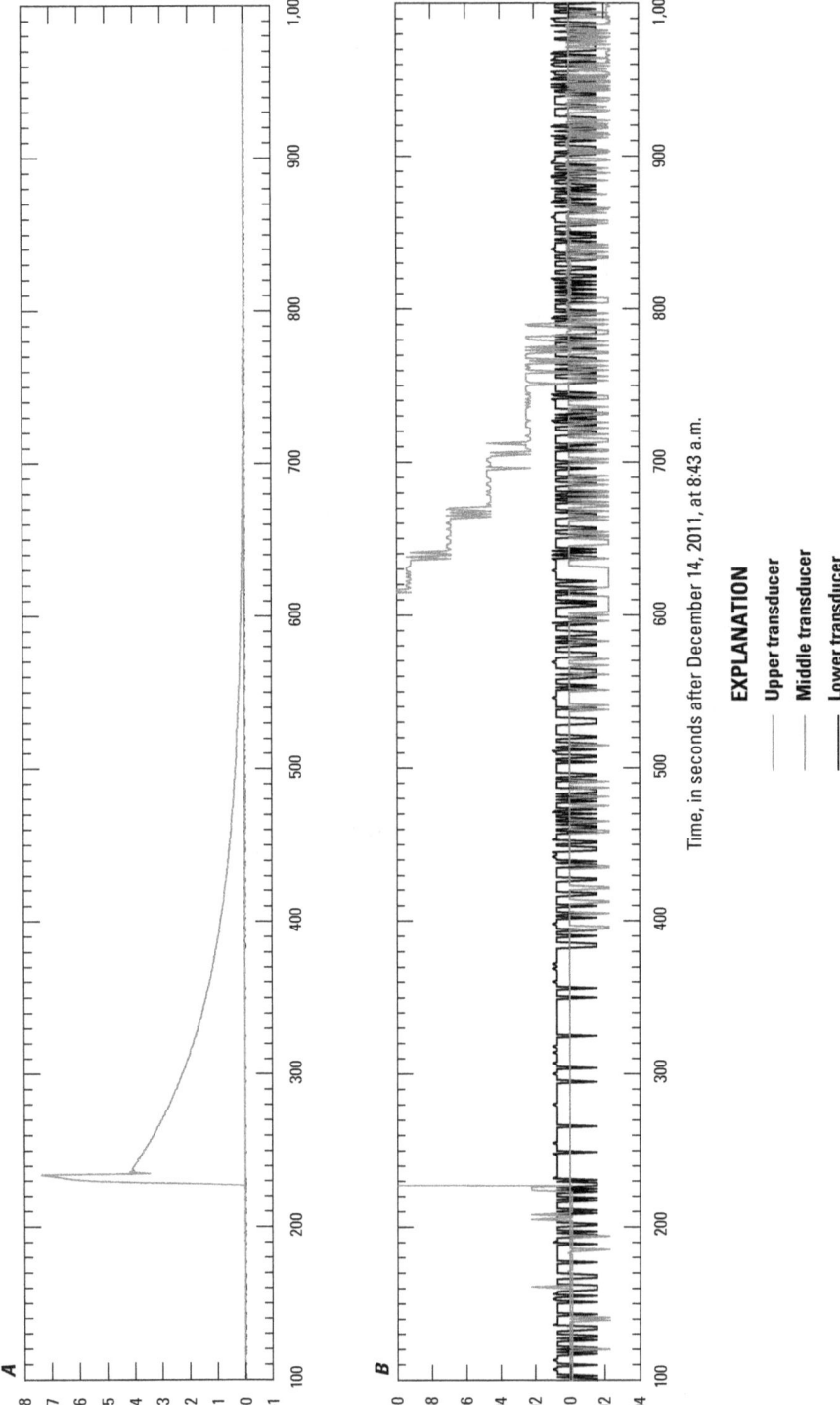

Figure 1–5. Water-level response to inserting a slug into the water column connected to the packer-isolated interval at 642 to 648 feet below land surface, Lower Floridan confining unit, test hole 35Q069, Pooler, Georgia, December 14, 2011. Static water level was 46.1 feet below land surface. *A*, Complete vertical extent of water-level response. *B*, Larger vertical scale of water-level response.

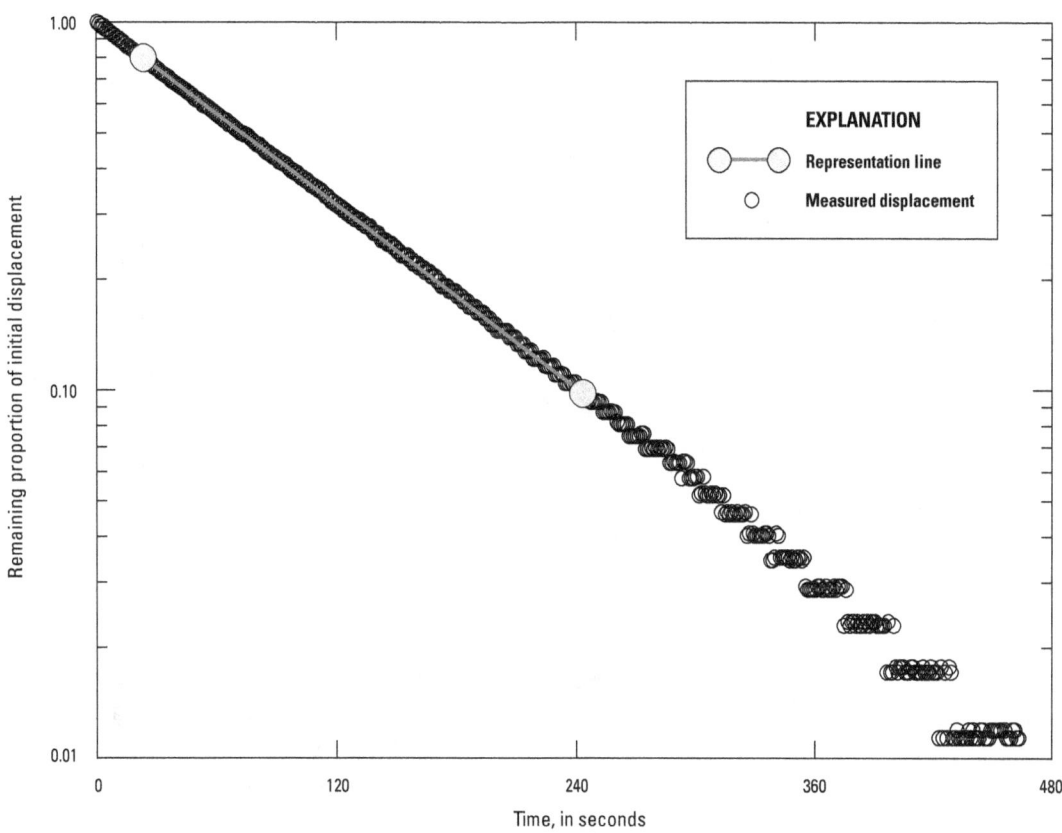

Figure 1–6.   Semi-log plot of water-level recovery from inserting slug into the water column connected to the packer-isolated interval at 642 to 648 feet below land surface, Lower Floridan confining unit, test hole 35Q069, Pooler, Georgia, December 14, 2011. Representation line was used to determine the horizontal hydraulic conductivity of the packer-isolated interval, using the Bouwer and Rice (1976) method. Modified from a graph on a spreadsheet from Halford and Kuniansky (2002).

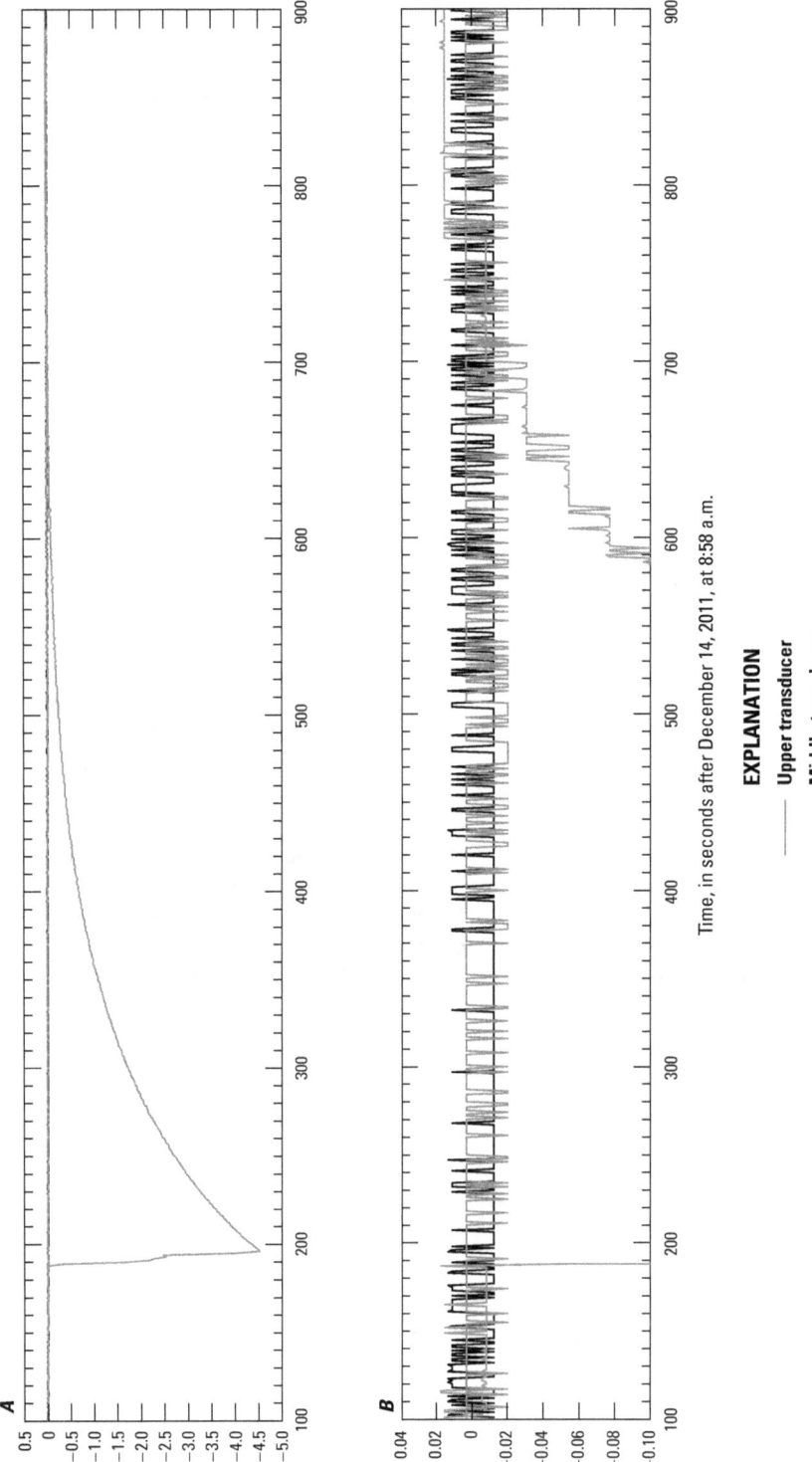

Figure 1–7.   Water-level response to pulling slug from the water column connected to the packer-isolated interval at 642 to 648 feet below land surface, Lower Floridan confining unit, test hole 35Q069, Pooler, Georgia, December 14, 2011. Static water level was 46.1 feet below land surface. *A*, Complete vertical extent of water-level response. *B*, Larger vertical scale of water-level response.

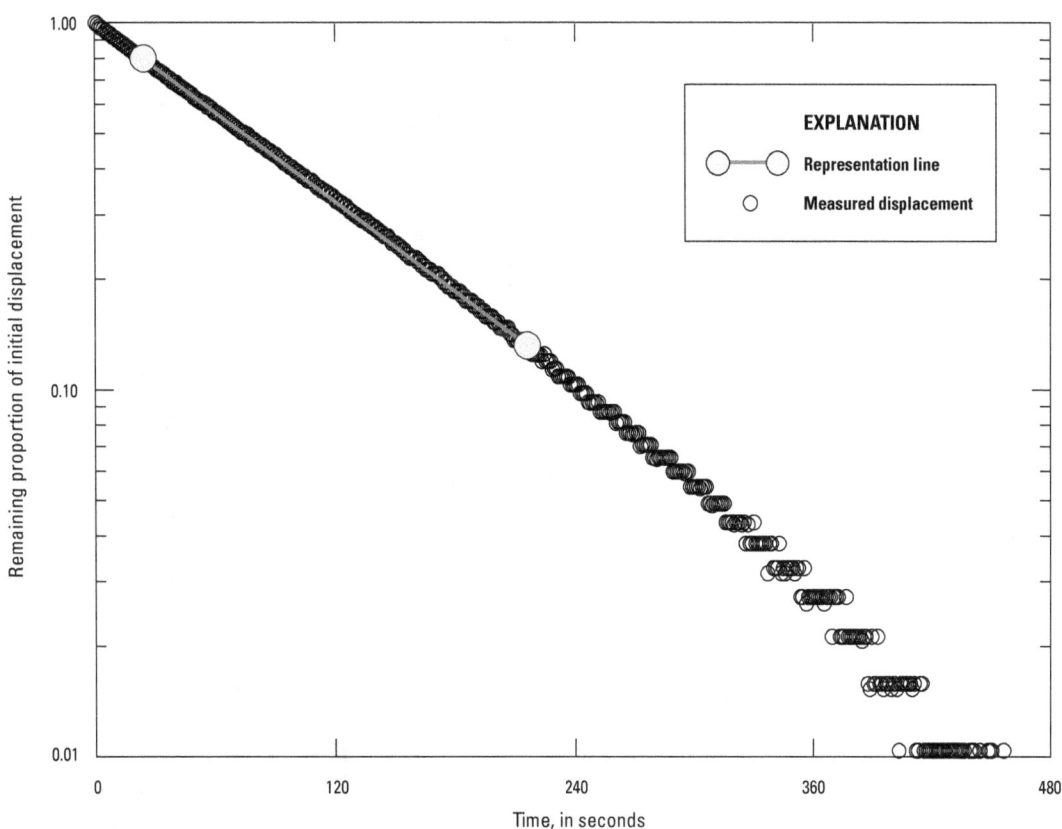

Figure 1–8.   Semi-log plot of water-level recovery from pulling slug from the water column connected to the packer-isolated interval at 642 to 648 feet below land surface, Lower Floridan confining unit, test hole 35Q069, Pooler, Georgia, December 14, 2011. Representation line was used to determine the horizontal hydraulic conductivity of the packer-isolated interval, using the Bouwer and Rice (1976) method. Modified from a graph on a spreadsheet from Halford and Kuniansky (2002).

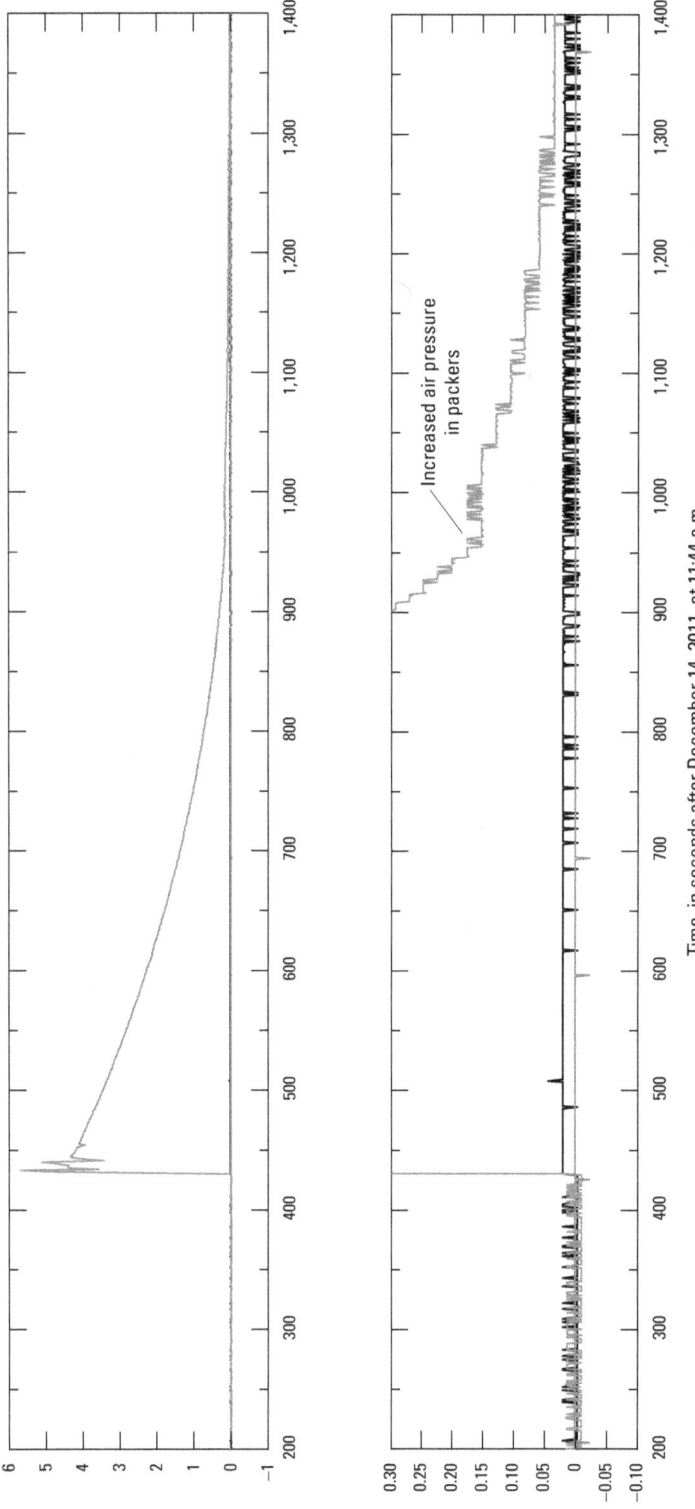

Figure 1–9.   Water-level response to inserting slug into the water column connected to the packer-isolated interval at 690 to 696 feet below land surface, Lower Floridan confining unit, test hole 35Q069, Pooler, Georgia, December 14, 2011. Static water level was 46.1 feet below land surface. *A*, Complete vertical extent of water-level response. *B*, Larger vertical scale of water-level response.

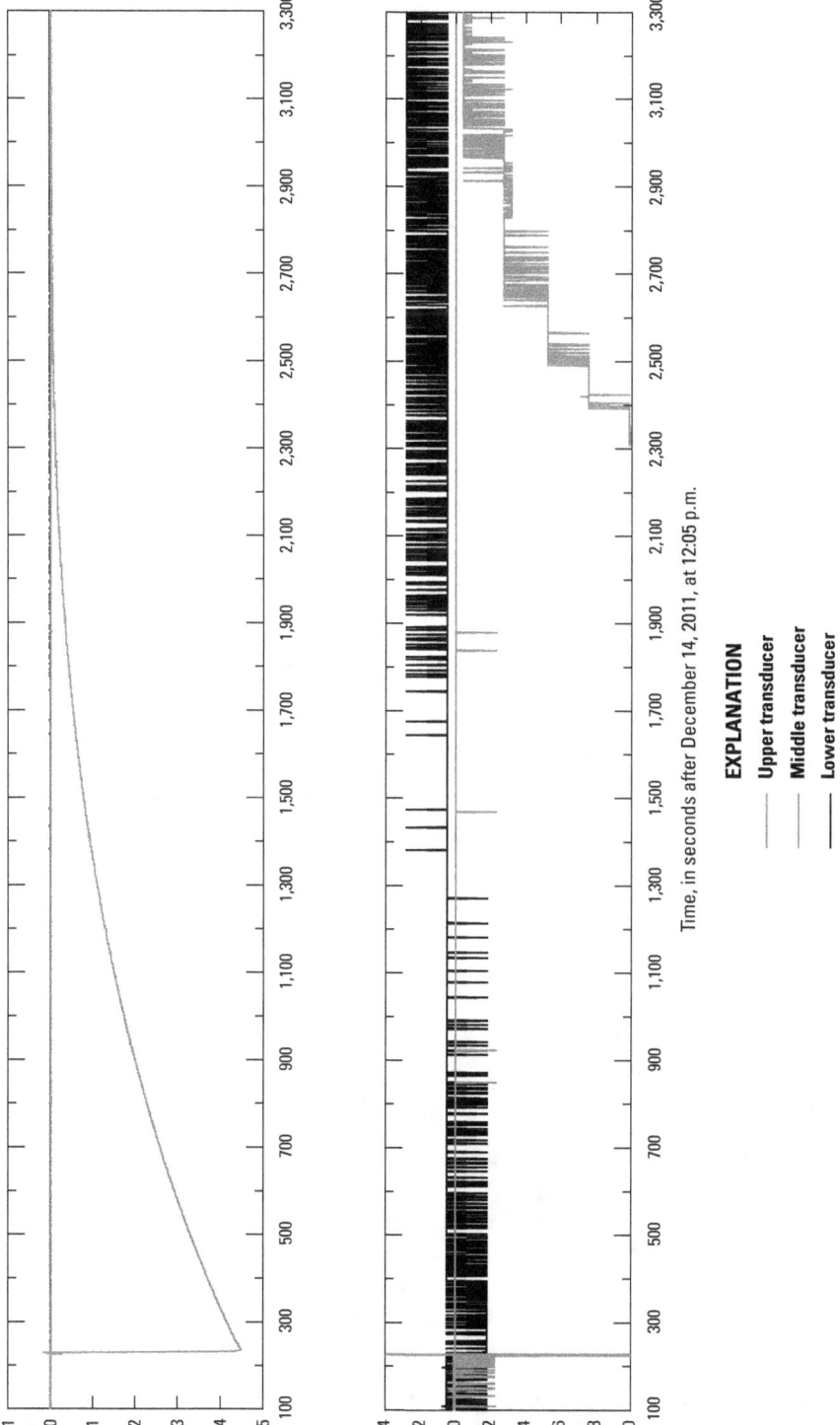

Figure 1–10. Water-level response to pulling slug from the water column connected to the packer-isolated interval at 690 to 696 feet below land surface, Lower Floridan confining unit, test hole 35Q069, Pooler, Georgia, December 14, 2011. Static water level was 46.1 feet below land surface. *A*, Complete vertical extent of water-level response. *B*, Larger vertical scale of water-level response.

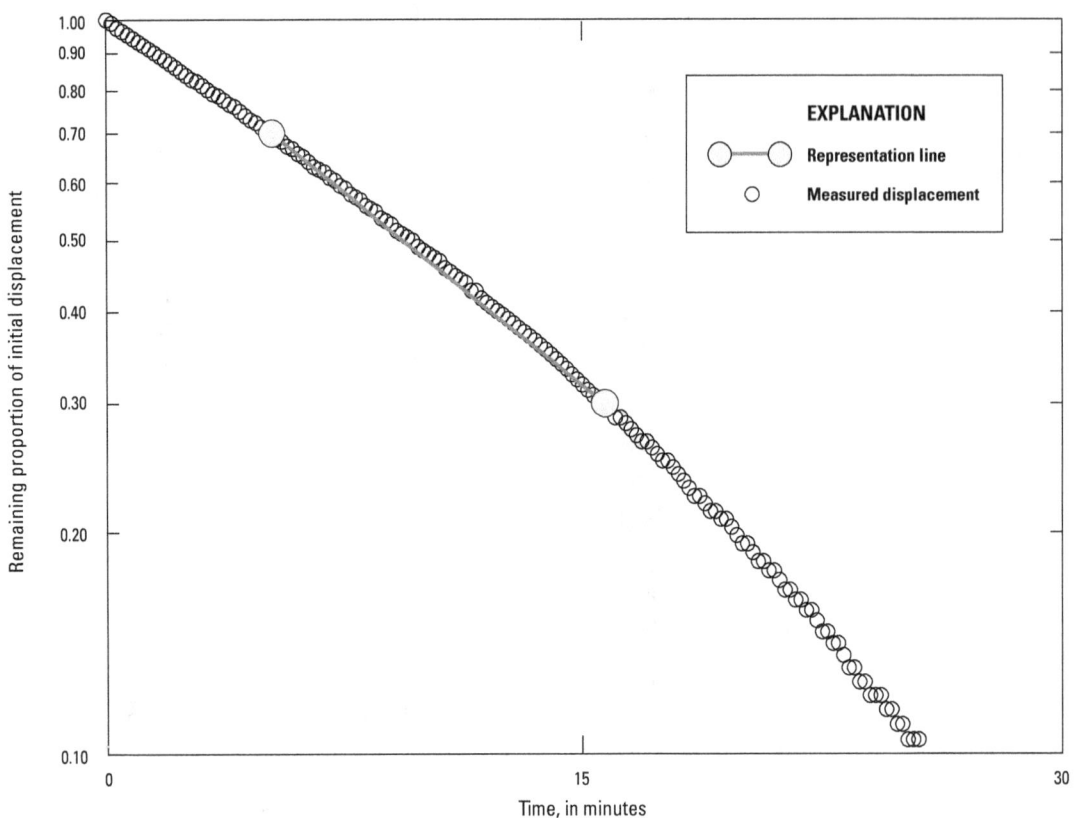

Figure 1–11.   Semi-log plot of water-level recovery from pulling slug from the water column connected to the packer-isolated interval at 690 to 696 feet below land surface, Lower Floridan confining unit, test hole 35Q069, Pooler, Georgia, December 14, 2011. Representation line was used to determine the horizontal hydraulic conductivity of the packer-isolated interval, using the Bouwer and Rice (1976) method. Modified from a graph on a spreadsheet from Halford and Kuniansky (2002).

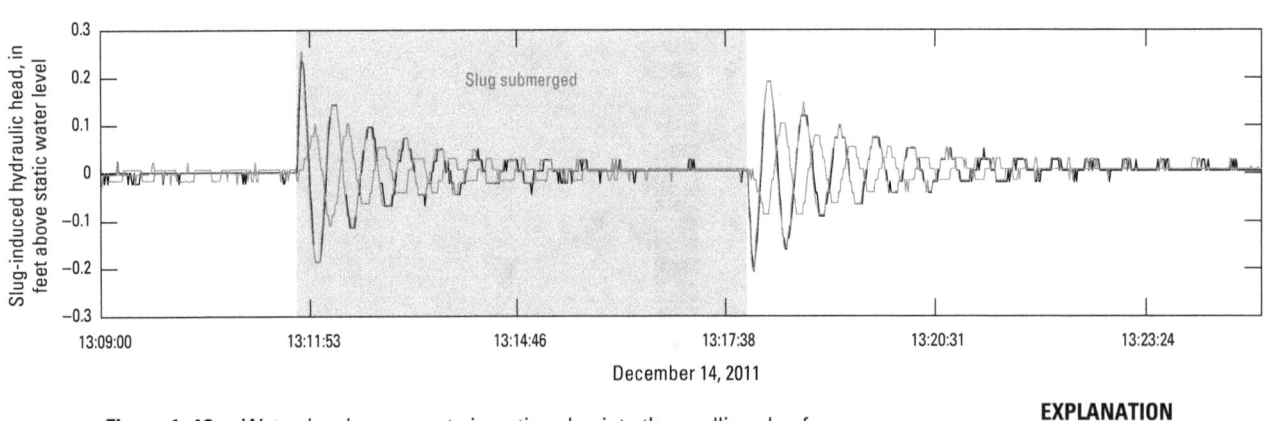

Figure 1–12.   Water-level response to inserting slug into then pulling slug from the water column connected to the Floridan aquifer system from 339 to 1,130 feet below land surface, test hole 35Q069, Pooler, Georgia, December 14, 2011. Static water level was 46.0 feet below land surface.

EXPLANATION

Upper transducer
Middle transducer
Lower transducer

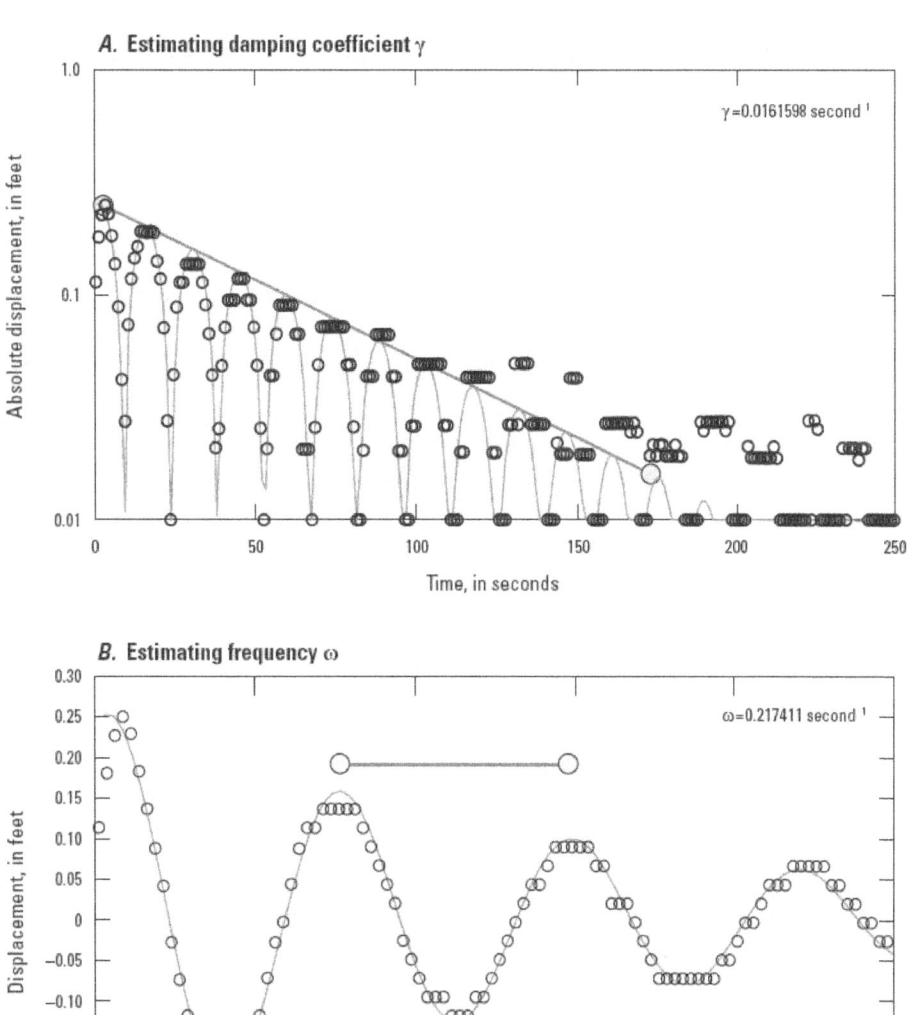

A. Estimating damping coefficient γ

$\gamma = 0.0161598$ second⁻¹

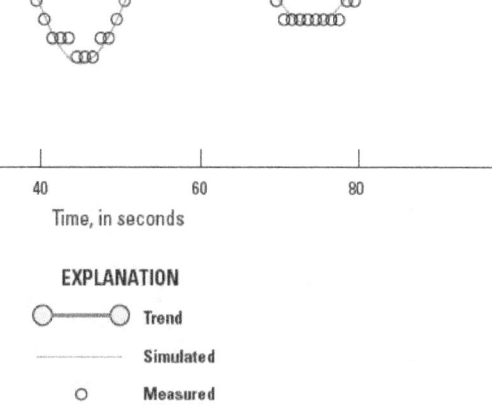

B. Estimating frequency ω

$\omega = 0.217411$ second⁻¹

EXPLANATION

○——○ Trend

······· Simulated

○ Measured

Figure 1–13. Measured and simulated water levels for inserting slug into water column open to the Floridan aquifer system from 339 to 1,130 feet below land surface, Pooler, Georgia, December 14, 2011. *A,* Log of absolute value of displacement as a function of time, used to determine damping coefficient γ. *B,* Water displacement as a function of time, used to determine oscillation frequency ω. Simulation is based on van der Kamp (1976) method. Graphs are modified from spreadsheet from Halford and Kuniansky (2002).

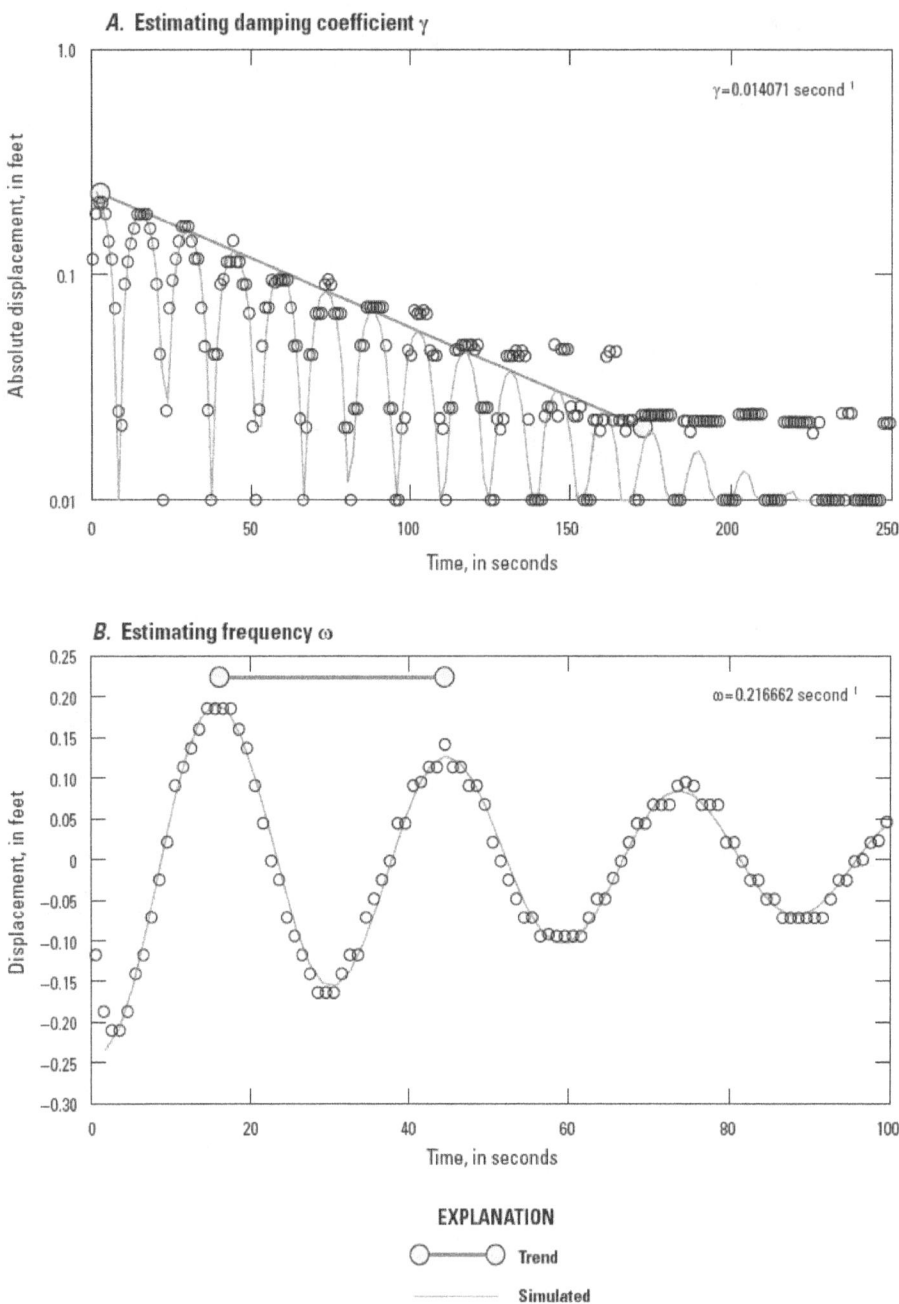

Figure 1–14. Measured and simulated water levels for pulling slug from water column open to the Floridan aquifer system from 339 to 1,130 feet below land surface, Pooler, Georgia, December 14, 2011. *A*, Log of absolute value of displacement as a function of time, used to determine damping coefficient *γ*. *B*, Water displacement as a function of time, used to determine oscillation frequency *ω*. Simulation is based on van der Kamp (1976) method. Graphs are modified from spreadsheet from Halford and Kuniansky (2002).

Appendix 2. Aquifer Tests, Pooler, Georgia, March and April 2012

Aquifer tests were performed at the Pooler test site to estimate the transmissivity of the Upper and Lower Floridan aquifers and to determine the effects of pumping one aquifer on water levels in the other aquifer. A 24-hour aquifer test was performed during March 27–28, 2012, in well 35Q070 open to the Upper Floridan aquifer (UFA), and a 72-hour aquifer test was performed during April 16–19, 2012, in well 35Q069 open to the Lower Floridan aquifer (LFA).

Wells 35Q069 and 35Q070 were monitored at least from March 26 to May 8, 2012. This timeframe included a pre-test period of slightly more than 16 hours, a 24-hour aquifer test period in the UFA, a failed 72-hour aquifer test period in the LFA, a 72-hour aquifer test period in the LFA, and a post-test period of at least 10 days. Numerous disruptions occurred in the continuous water-level record during the study period (table 2–1, fig. 9), caused by pumping events in well 35Q069 and data gaps in both wells due to temporary removal of transducers.

Drawdown Estimation

Drawdown for the two wells was estimated by using a procedure developed by Halford (2006a) to filter water-level data for effects of barometric pressure, earth and ocean tides, and long-term trends. Time series related to known influences on water levels are used as explanatory variables to synthesize a water-level time series (fig. 2–1). The synthetic water levels closely approximate (are matched to) measured water levels during periods that are unaffected by an aquifer test (hereinafter referred to as the "fitting period"; fig. 2–2). Measured water level during the fitting period is used to determine phase shifts and multipliers of explanatory variables that will synthesize a water level that matches the measured water levels. Differences between the synthetic and measured water levels are minimized using a sum-of-squares objective function. The synthetic water level is further adjusted using the offset and slope with respect to time to match the synthetic water level with the measured water level just before the start of the aquifer test and just after full recovery from the aquifer test. The difference between synthetic and measured water level during the aquifer test and recovery represents drawdown induced by pumping; the synthetic water levels represent nonpumping water levels.

Four drawdown conditions were evaluated during the aquifer tests: (1) response of the UFA well to the 24-hour aquifer test in the UFA (fig. 2–3); (2) response of the LFA well to the 24-hour aquifer test in the UFA (fig. 2–4);

Table 2–1. Disruptions in continuous monitoring of natural water-level during the study period, Pooler, Georgia, February through April 2012.

[NA, not applicable; gal/min, gallon per minute; UFA, Upper Floridan aquifer; LFA, Lower Floridan aquifer]

Well	Start time	End time	Event description
	(Month day time)		
35Q069	February 23 1 p.m.	NA	Start monitoring
35Q069	March 4 3:30 p.m.	March 5 0:45 a.m.	Unnatural spike
35Q069	March 20 6:15 p.m.	March 26 4:45 p.m.	Unnatural daily fluctuations characteristic of a clogged vent
35Q070	March 26 4:30 p.m.	NA	Start monitoring
Both	March 27 9:00:03 a.m.	March 28 9:05:03 a.m.	24-hour aquifer test (pumping UFA well 35Q070 at 285 gal/min)
Both	March 28 9:05:03 a.m.	March 31 12 a.m.	24-hour aquifer test recovery
35Q070	April 2 11:50 a.m.	April 4 8:15 a.m.	No data
36Q069	April 3 1 p.m.	April 4 5:30 p.m.	No data
36Q069	April 4 6:30 p.m.	April 4 7:15 p.m.	Pumping event in LFA well 35Q069 and recovery
36Q069	April 5 2:45 p.m.	April 5 8:15 p.m.	Pumping event in LFA well 35Q069 and recovery
36Q069	April 10 6 a.m.	April 10 11:36 a.m.	No data
Both	April 10 12:00:44 p.m.	April 10 3:20 p.m.	Failed 72-hour aquifer test (pumping LFA well 35Q069 at 1,200 gal/min)
Both	April 10 3:20 p.m.	April 11 12 p.m.	Failed 72-hour aquifer test recovery
36Q069	April 12 7 a.m.	April 13 12 p.m.	Pumping event in LFA well 35Q069 and recovery
Both	April 16 12:00:04 p.m.	April 19 12:05:10 p.m.	72-hour aquifer test (pumping LFA well 35Q069 at 783 gal/min)
Both	April 19 12:05:10 p.m.	April 28 12 a.m.	72-hour aquifer test recovery
Both	NA	May 8 2 p.m.	End of monitoring

(3) response of the LFA well to the 72-hour aquifer test in the LFA (fig. 2–5); and (4) response of the UFA well to the 72-hour aquifer test in the LFA (fig. 2–6).

Barometric pressure and gravity along with water levels from four background wells (36Q020, 33R045, 35P110, and 35P125) and ocean tides from Savannah River stream gage (station number 02198980) were used as explanatory time-series components to filter out nonpumping influences during the aquifer tests (locations shown in fig. 1). Barometric-pressure data were compiled from hourly barometric-pressure data from National Oceanic and Atmospheric Administration (NOAA) weather station 097847, located about 6 miles northeast of the Pooler test site (097847 location shown in fig. 1). Ocean-tide data were provided from the Savannah River stream gage 02198980, located at the mouth of the Savannah River where it enters the Atlantic Ocean. Gravity is used as a surrogate for earth tides and is expressed as microgravity deviation from average. The time-series spreadsheet from Halford (2006a) generates gravity time series from entered values of latitude, longitude, and land-surface altitude. Background wells are far enough away from the Pooler test site so that they do not show a response (zero drawdown) to the aquifer tests but have a response to nonaquifer-test influences that are similar to those of the test wells. Background wells 36Q020 and 35P110 are open to the UFA; background wells 33R045 and 35P125 are open to the LFA. Background well 36Q020 is strongly affected by earth tides. Water levels in background wells and barometric pressure, gravity, and ocean tides are shown in figure 2–1.

The synthetic water level for a well w during a given period j at a given time t, is:

$$Syn_{(w,j,t)} = C_{(w,j)} + m_{(w,j)} \times (t - t_0) + \sum_{i=1}^{n} a_i \times V_{(i,t+\varphi_i)} \qquad (3)$$

where

$Syn_{(w,j,t)}$ is the synthetic water level of well w during time period j at time t, in feet above datum;

$C_{(w,j)}$ is the offset for well w during time period j, in feet above datum;

$m_{(w,j)}$ is the linear water-level trend with respect to time (slope) for well w during time period j, in feet per day;

t_0 is an arbitrary reference time; $(t - t_0)$ in days;

n is the number of explanatory time-series components, ranging from 3 to 12 in this study;

a_i is the amplitude of the multiplier for the ith explanatory time-series component, in feet per unit of the explanatory time-series component;

$V_{(i,t+\varphi i)}$ is the value of the ith explanatory time-series component at time $t + \varphi_i$; and

φ_i is the phase shift of the ith explanatory time-series component, in days.

The parameters a_i and φ_i are adjusted for each explanatory time-series component and $m_{(w,j)}$ and $C_{(w,j)}$ are applied to the synthetic water-level time series. The time period j is either the specific fitting period or the period of time from before to after the aquifer test and recovery. The data set that makes up j is the time period plus a phase shift buffer before the start and after the end of j. The buffers allow the start and end of the explanatory time series to shift (φ_i) beyond the bounds of j. For a given combination of explanatory time-series components and fitting period, the time-series spreadsheet is used to adjust the amplitude of multipliers, phase shifts, slope with respect to time, and offset to fit the synthetic water level to the measured water level during the fitting period. The spreadsheet and supporting program are used to fit the synthetic water level to the measured water level by minimizing their differences with a sum-of-squares objective function.

Numerous disruptions in the water-level data of the test wells limit the times that can be used as fitting periods (table 2–1). Two fitting periods ultimately were used: an early fitting period beginning midnight on March 31, 2012, and ending at 11 a m. on April 2, 2012; and a late fitting period beginning midnight on April 28, 2012, and ending at 9 a m. on May 6, 2012. Halford (2006a) indicates that the fitting period ideally would include the period immediately antecedent of the aquifer test and should extend at least four times longer than the prediction period. Given that the prediction period should include the pumping and recovery periods, the required fitting period would extend about 2 weeks for the 24-hour aquifer test and about 4 weeks for the 72-hour aquifer test. Unfortunately, the available fitting periods could not extend as long as required; therefore, some nonaquifer-test influence was not removed from the water-level data.

For each well-aquifer-test situation, a combination of explanatory time series and a fitting period were selected to best filter nonpumping influences from the water-level data. The process begins by selecting a candidate combination from the pool of available explanatory time-series components. The amplitudes of the multipliers and phase shifts are adjusted for each explanatory time series to minimize the sum-of-squares objective. If the synthesized water level for a candidate combination of explanatory time series matches closely to the selected fitting period (fig. 2–2), the amplitudes of multipliers and phase shifts used to match the fitting period are used to synthesize water levels from several hours before the aquifer test to about a day after the aquifer-test recovery. The offset and slope with respect to time were adjusted so that the synthetic water level matched the measured water for the time just before the aquifer test and just after the aquifer-test recovery. Different combinations of explanatory time series were used with both fitting periods to find the best match before and after the aquifer test and recovery (best matches shown in figs. 2–3 to 2–6). For the best match of synthetic water level with measured water level before the aquifer test and after the aquifer-test recovery, the synthetic water level during the aquifer test was taken as the water level not being affected by the aquifer test. The drawdown was the synthetic water level minus the measured water level.

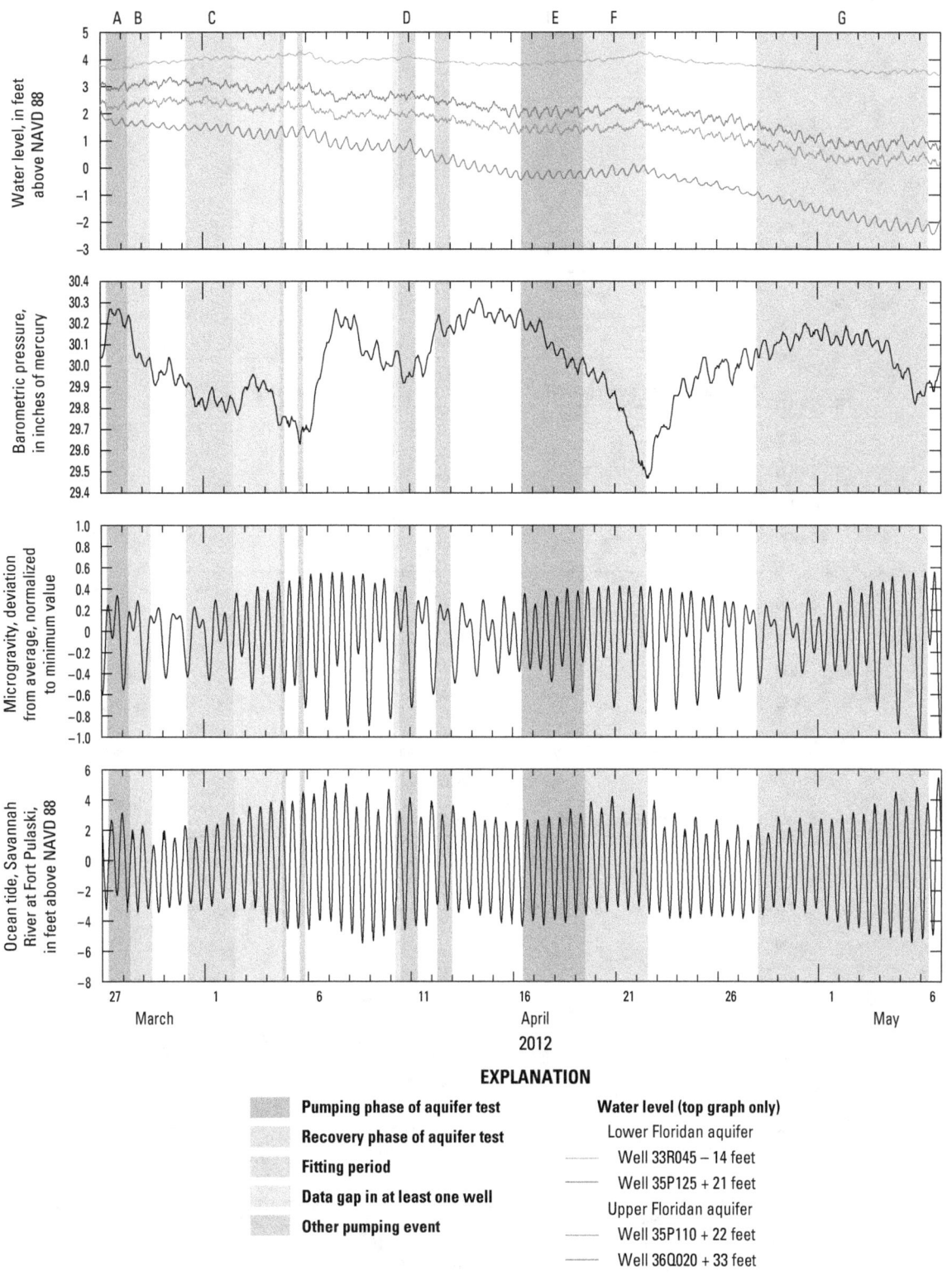

Figure 2–1.   Water-level fluctuations in background wells and fluctuations in barometric pressure, gravity, and ocean tides used in the aquifer tests, Pooler, Georgia, March 27–May 6, 2012. *A*, Pumping phase of Upper Floridan 24-hour aquifer test. *B*, Recovery phase of Upper Floridan 24-hour aquifer test. *C*, Early fitting period. *D*, Failed Lower Floridan aquifer test. *E*, Pumping phase of Lower Floridan 72-hour aquifer test. *F*, Recovery phase of Lower Floridan 72-hour aquifer test. *G*, Late fitting period. The fitting period is used to describe the match (or fit) of synthetic water levels to those that were measured.

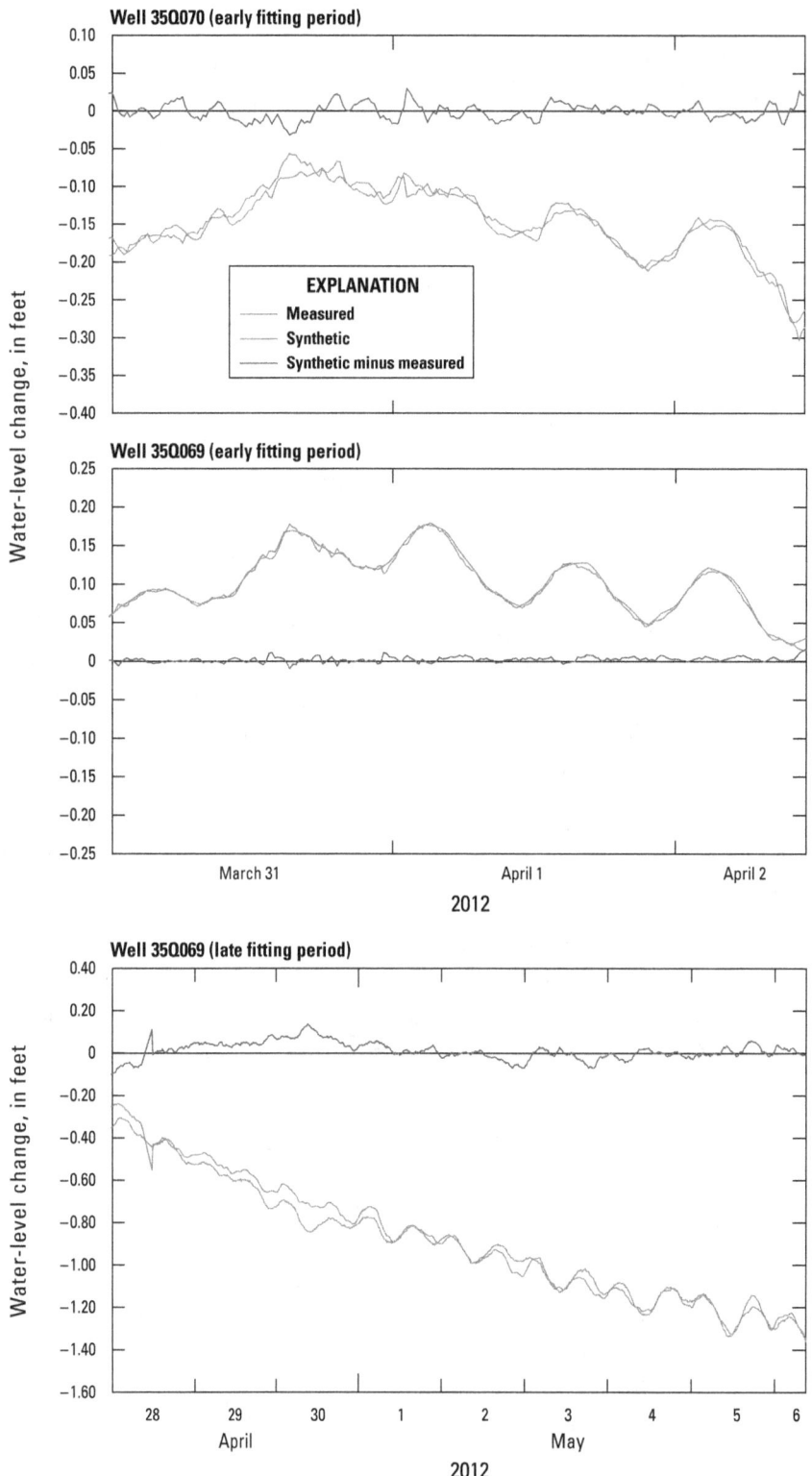

Figure 2–2. Fit of synthetic water levels to measured water levels for wells 35Q070 and 35Q069, Pooler, Georgia, April 28–May 6, 2012. The early fitting period was used to estimate drawdown in well 35Q070 in response to the 24- and 72-hour aquifer tests and in well 35Q069 in response to the 24-hour aquifer test. The late fitting period was used to estimate the drawdown in well 35Q069 in response to the 72-hour aquifer test.

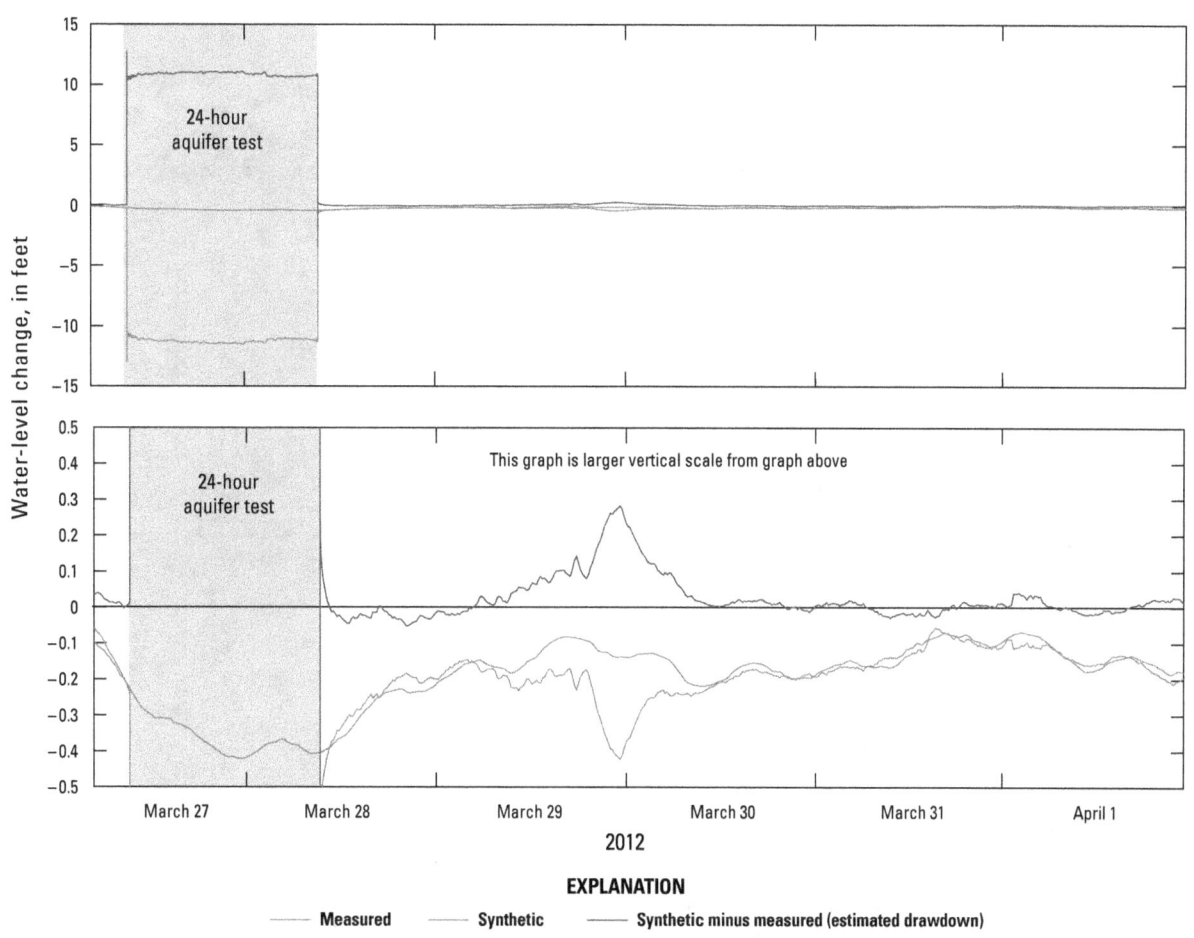

Figure 2–3.   Measured and synthetic water levels and estimated drawdown in Upper Floridan aquifer well 35Q070 during the 24-hour aquifer test performed in the well, Pooler, Georgia, March 27–April 1, 2012.

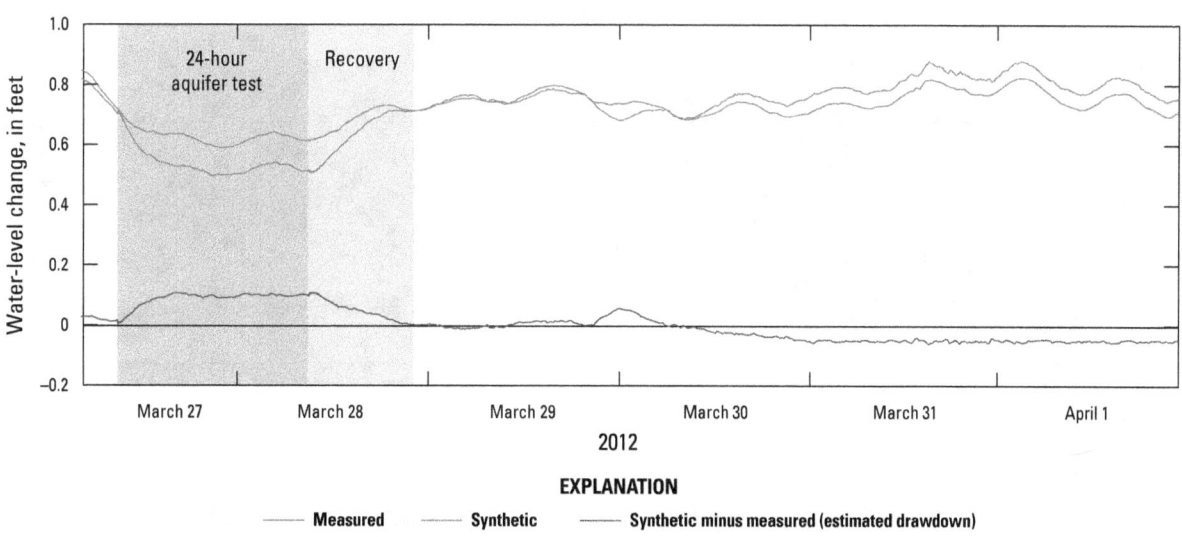

Figure 2–4.   Measured and synthetic water levels and estimated drawdown in Lower Floridan aquifer well 35Q069 during the 24-hour aquifer test performed in the Upper Floridan aquifer well 35Q070, Pooler, Georgia, March 27–April 1, 2012.

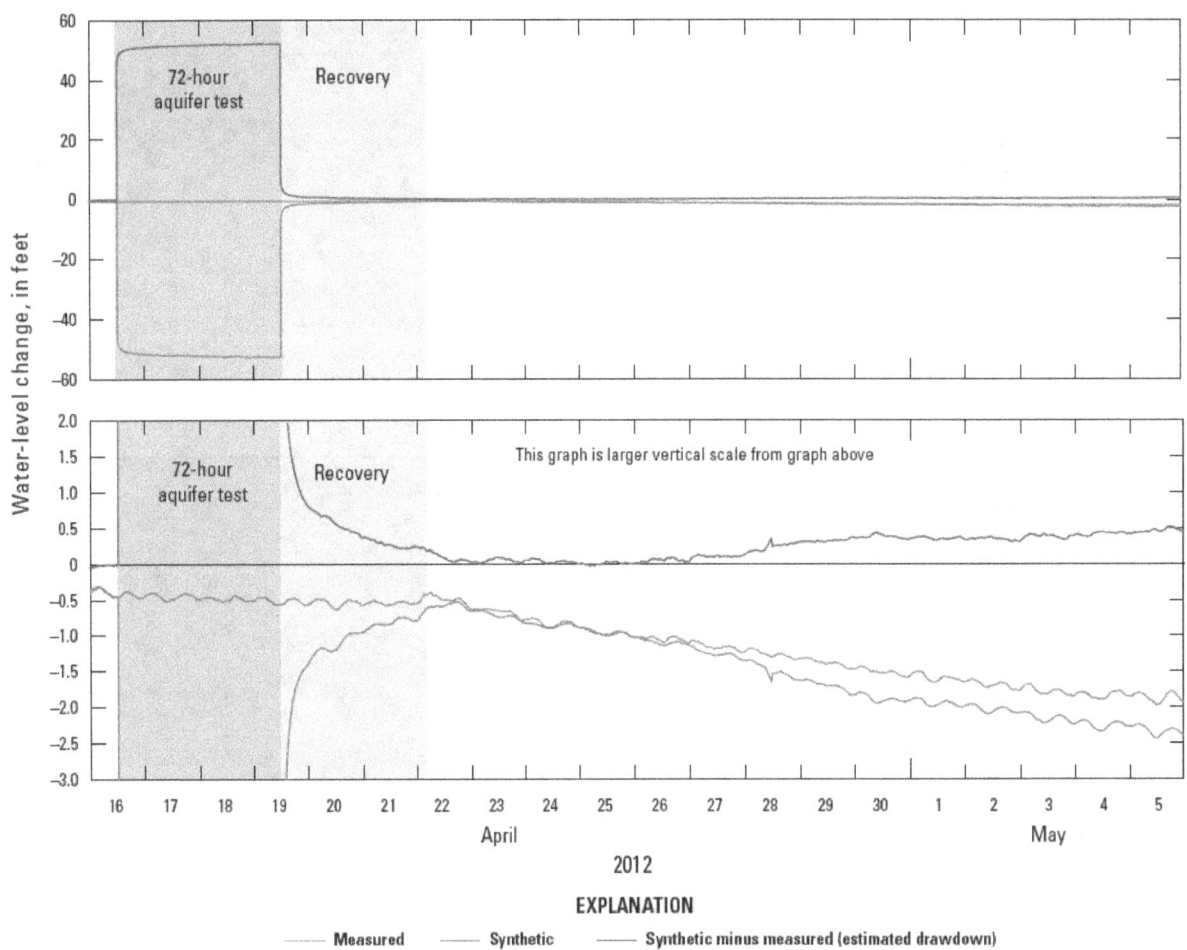

Figure 2–5. Measured and synthetic water levels and estimated drawdown in Lower Floridan aquifer well 35Q069 during the 72-hour aquifer test performed in the well, Pooler, Georgia, April 16–May 5, 2012.

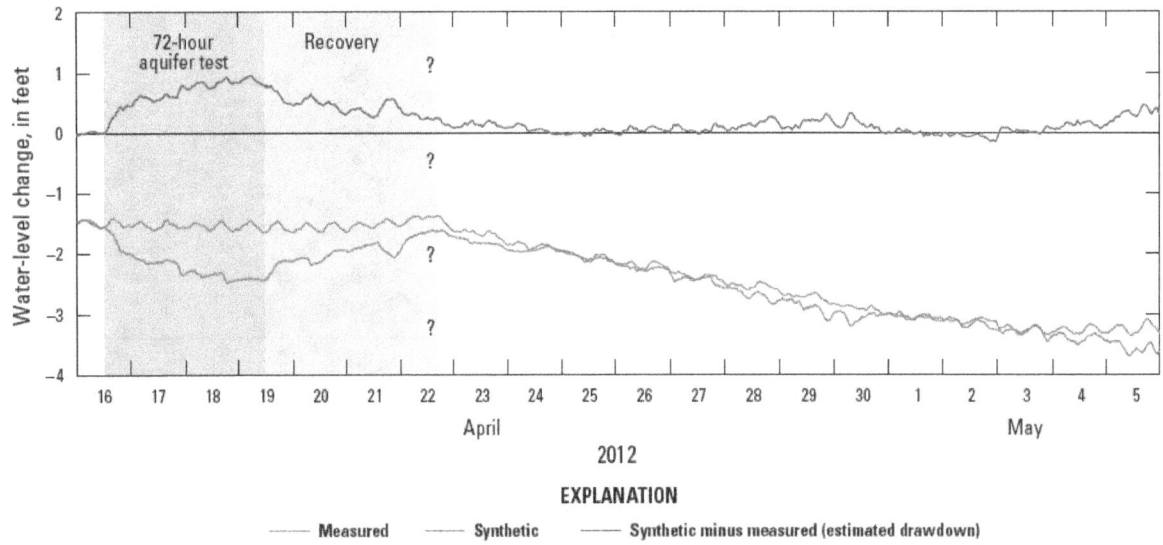

Figure 2–6. Measured and synthetic water levels and estimated drawdown in Upper Floridan aquifer well 35Q070 during the 72-hour aquifer test performed in the Lower Floridan aquifer well 35Q069, Pooler, Georgia, April 16–May 5, 2012.

24-Hour Aquifer Test

Pumping for the 24-hour aquifer test at UFA well 35Q070 began at 9:00 a.m. on March 27, 2012, and continued for 24 hours and 5 minutes until 9:05 a.m. on March 28, 2012 (fig. 2–3). The well was pumped at a rate that varied slightly (table 2–2). The pumping rate for the first 15 hours was 285 gallons per minute (gal/min). Starting at midnight on March 28, 2012, the rate decreased to about 281 gal/min and continued to decrease to slightly less than 280 gal/min. From early morning on March 28, 2012, until pumping was stopped, the discharge rate slowly increased back to 285 gal/min.

Table 2–2.   Pumping rate through the 24-hour aquifer test in well 35Q070, Pooler, Georgia, March 27–28, 2012.

Time of start of change in pump rate (Month day time)	Measured pump rate (gallons per minute)
March 26 4:30 p.m.	0
March 27 9:00:03 a.m.	287
March 27 9:15 a.m.	285
March 28 12 a.m.	281
March 28 3 a.m.	280
March 28 7 a.m.	281
March 28 8 a.m.	283
March 28 8:45 a.m.	285
March 28 9:05:03 a.m.	0

Drawdown Response

In response to aquifer-test pumping, the water level in well 35Q070 declined by 10.4 feet (ft) almost instantaneously, exhibiting random water-level fluctuations (noise) in the drawdown signal (fig. 2–7). In response to the end of the 24-hour pumping period, the water level rose to near zero drawdown almost instantaneously with some noise that diminished after about 100 seconds. The near-instantaneous decline in water level represents the turbulent component of drawdown (Batu, 1998). The laminar component of drawdown is a consistent increase in drawdown with time, which is used to estimate the transmissivity of the UFA.

Filtering the aquifer-test water-level response for barometric-pressure change seemed to best filter out erratic water-level fluctuations; however, no combination of explanatory time-series components was effective in filtering out much nonaquifer-test influence on aquifer-test water levels. The drawdown time series resulting from applying several different filters to mitigate the nonaquifer-test noise produced similar results (fig. 2–8).

During the 15 minutes to 15 hours aquifer testing, drawdown consistently increased with log time (constant log cycle; fig. 2–9). At midnight on March 28, 2012, drawdown decreased. At this time, the discharge rate suddenly decreased from 285 to 281 gal/min, which might have caused the decrease in drawdown.

The decline in water level (caused by aquifer-test and other influences) during the 24-hour aquifer test at UFA pumped well 35Q070 equaled about 11.2 ft, from 41.7 to about 52.9 ft below land surface. The water-level decline was obscured by barometric-pressure change. Adjusting the drawdown for barometric-pressure change indicates a corrected drawdown, in response to the aquifer test, of about 11.1 ft as of midnight March 28, 2012, (fig. 2–3).

The water-level decline in LFA well 35Q069, in response to the 24-hour aquifer test in UFA pumped well 35Q070, was obscured by barometric-pressure change, earth tides, and ocean tides. Twelve explanatory time-series components, including all four background wells and the fitting period from midnight on March 31, 2012, to 11:00 a.m. on April 2, 2012, were used to filter out nonaquifer-test influences. Synthetic water levels matched well with measured water levels before and after the 24-hour pumping period and recovery (fig. 2–4). Drawdown began to increase approximately 20 minutes after the start of the 24-hour aquifer test. From about 20 minutes to about 8 hours after the start of the 24-hour aquifer test, drawdown increased from 0 to 0.09 ft. From 8 hours after the start of the aquifer test to the end of the pumping period, drawdown fluctuated slightly, but stayed at about 0.09 ft.

Aquifer-Test Analysis

The Cooper-Jacob, straight-line method (Cooper and Jacob, 1946) was applied to the drawdown data from pumped well 35Q070 to determine the transmissivity of the UFA (fig. 2–9). Using the laminar drawdown (as described by Batu, 1998) between 15 minutes and 15 hours after the start of the aquifer test, the transmissivity was estimated to be 30,000 feet squared per day (ft^2/d). The data points used to determine the log-cycle of drawdown consist of clusters on either end of the line, which created uncertainty in the selection of distinct analytical end points to the straight line and therefore the slope. Transmissivities for different fits to the drawdown data range from about 27,000 to 33,000 ft^2/d and round to 30,000 ft^2/d.

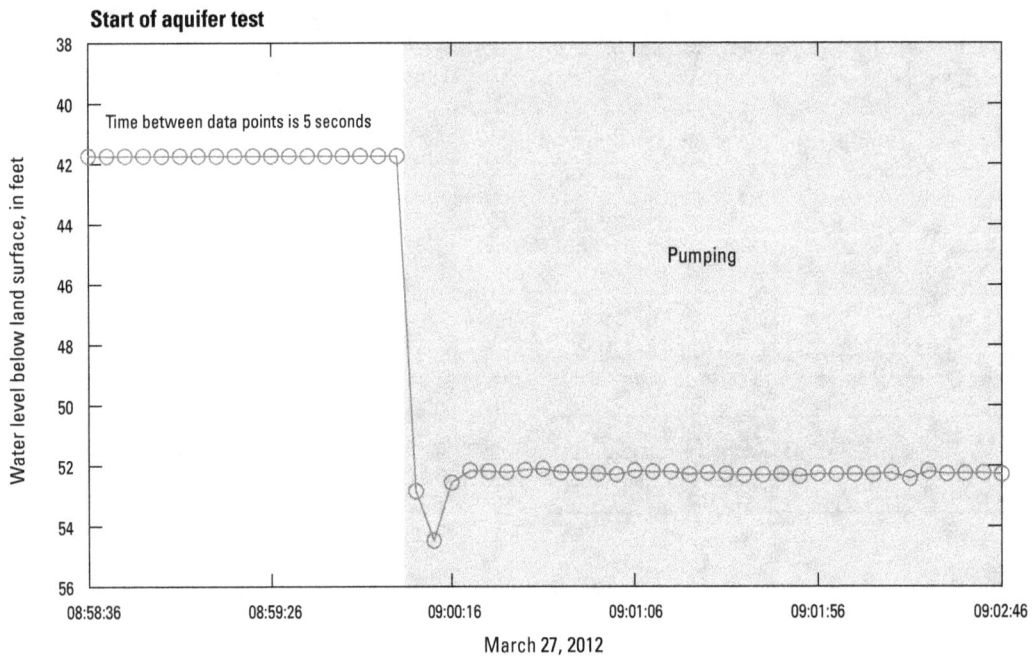

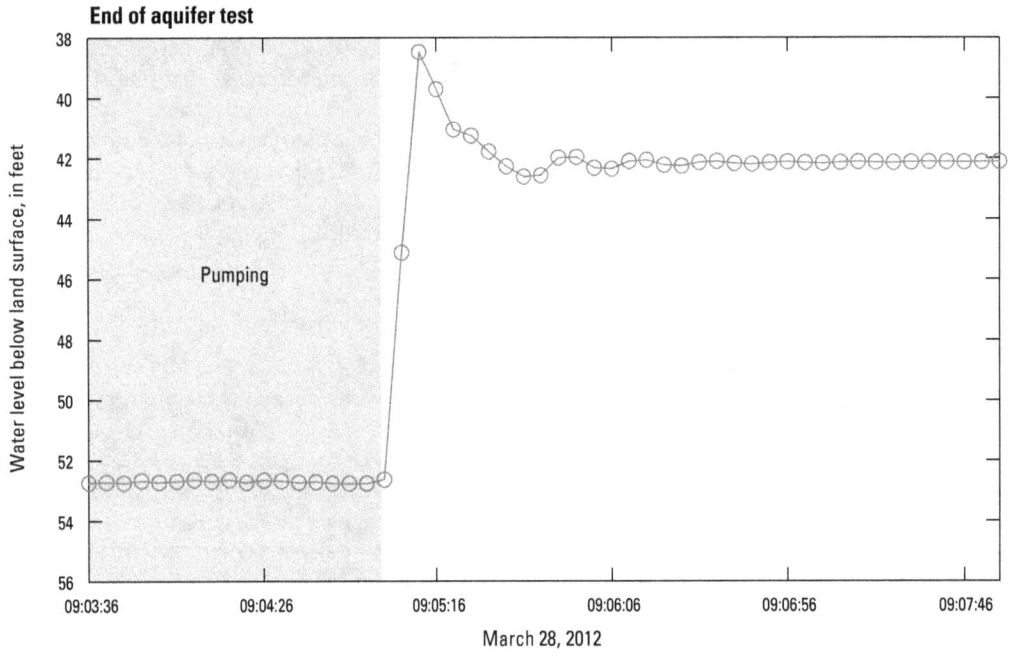

Figure 2–7.   Measured water levels at the start and end of the 24-hour aquifer test in Upper Floridan aquifer well 35Q070, Pooler, Georgia, March 27–28, 2012.

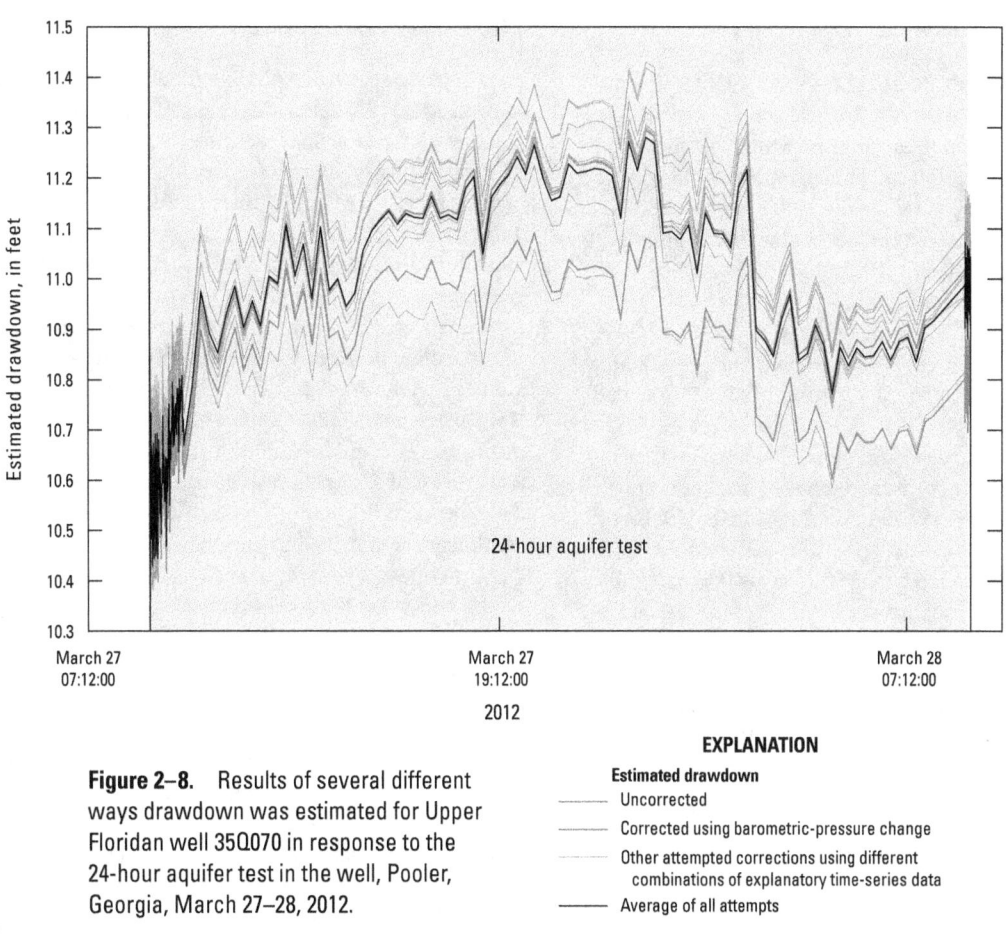

Figure 2–8. Results of several different ways drawdown was estimated for Upper Floridan well 35Q070 in response to the 24-hour aquifer test in the well, Pooler, Georgia, March 27–28, 2012.

EXPLANATION

Estimated drawdown

——— Uncorrected

——— Corrected using barometric-pressure change

——— Other attempted corrections using different combinations of explanatory time-series data

——— Average of all attempts

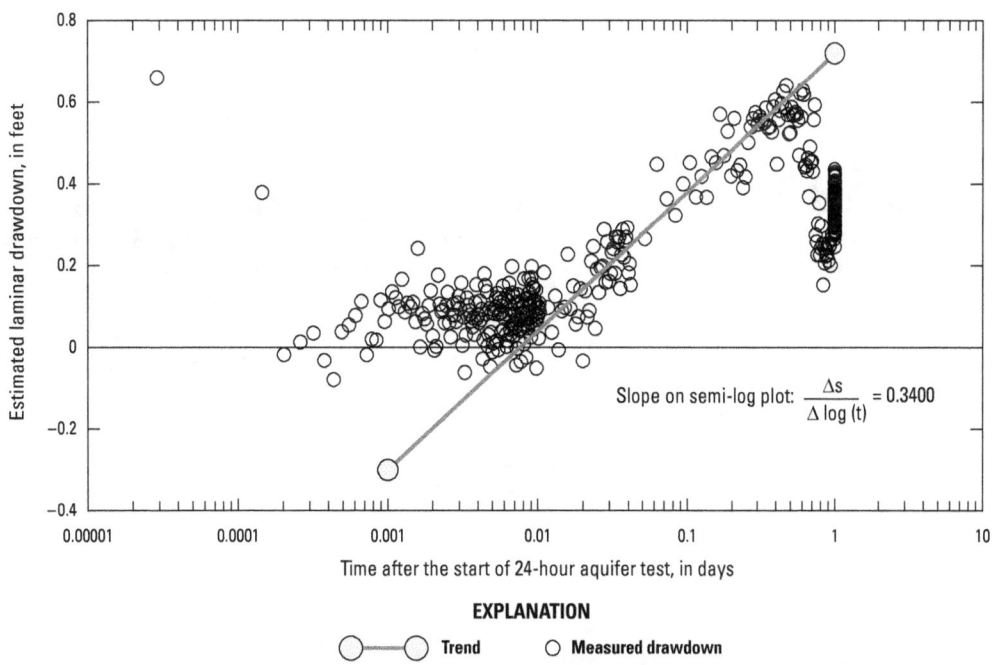

Slope on semi-log plot: $\dfrac{\Delta s}{\Delta \log (t)} = 0.3400$

EXPLANATION

◯——◯ Trend   ◯ Measured drawdown

Figure 2–9. Semi-log graph of drawdown during the 24-hour aquifer test in well 35Q070 at a pumping rate of about 285 gallons per minute, Pooler, Georgia, March 27–28, 2012.

72-Hour Aquifer Test

The 72-hour aquifer test at LFA well 35Q069 was first attempted at noon on April 10, 2012. The well was pumped at a rate of 1,200 gal/min, much greater than the planned pumping rate for the proposed public-supply well for the City of Pooler, Georgia. Because the discharge rate was too high, the aquifer test was discontinued 3 hours and 20 minutes after pumping had started. The well was then allowed 6 days to recover from the failed aquifer test.

The second aquifer-test pumping at LFA well 35Q069 began at noon on April 16, 2012, and continued for 72 hours and 5 minutes until 12:05 p.m. on April 19, 2012. The well was pumped at rates of 780 to 788 gal/min (table 2–3). The pump lacked a check value; therefore, when the aquifer-test pumping ceased, a slug of water siphoned back into the well. As a result, the water level shows a slug signature during the first 3 minutes of recovery (fig. 2–10). Within 3.5 days, the water level in pumped well 35Q069 was dominated by a long-term decline in background water level (figs. 9 and 2–5).

Table 2–3. Pumping rate through the 72-hour aquifer test in well 35Q0690, Pooler, Georgia, April 16–19, 2012.

Time of start of change in pump rate (Month day time)	Measured pump rate (gallons per minute)
April 12 1:30 p.m.	0
April 16 12:00:04 p.m.	788
April 16 12:55 p.m.	788
April 16 3 p.m.	783
April 16 6 p.m.	783
April 16 7 p.m.	780
April 19 12:05:09 p.m.	780
April 19 12:05:10 p.m.	0

Drawdown Response

Filtering caused negligible modification to the drawdown of well 35Q069 when it was pumped, because the magnitude of the drawdown signal (greater than 50 ft) obscured any signals corresponding to nonpumping influences, which probably ranged from a few hundredths of a foot to about 0.4 ft (fig. 9). Pre-test data indicated a minor (up to 0.1 ft) cyclic diurnal fluctuation in water level, which is characteristic of earth tides, and more substantial (up to 0.4 ft) fluctuations in water levels, which is a characteristic of barometric-pressure change.

Decline in water level (caused by aquifer-test and other influences) during the 72-hour aquifer test at pumped well 35Q069 equaled 52.0 ft, from 43.4 to 95.4 ft below land surface. Adjusting nonaquifer-test influences indicates a corrected drawdown of 51.7 ft in response to the aquifer test (fig. 2–5). Drawdown at the pumped well as a function of log (time) was nonlinear, as indicated by the continuously decreasing log cycle of drawdown with time (fig. 2–11). The nonlinear nature of drawdown on the semi-log plot precluded the viable use of an analytical method for estimating the transmissivity of the LFA. The rate of drawdown slightly increased 0.007 day (5 minutes) after the start of the aquifer test. No discharge-rate change was documented at the time. This slight change might be caused by lateral heterogeneity in the hydraulic properties of the Floridan aquifer system at the Pooler test site.

Water-level decline in UFA well 35Q070 in response to the 72-hour aquifer test in LFA pumped well 35Q069 was obscured by daily fluctuations and long-term water-level trends. Water-level decline in UFA well 35Q070 (caused by aquifer test and other influences) during the 72-hour aquifer test at the LFA pumped well was 0.8 ft, from 43.1 to 43.9 ft below land surface (−24.1 to −24.9 ft above NAVD 88; fig. 9). Adjusting for nonaquifer-test influences indicates a corrected drawdown of 0.9 ft (fig. 2–6). Drawdown was obscured mostly by a change in long-term water-level trend on April 23, 2012, from relatively unchanging to declining and by daily fluctuations with an amplitude of about 0.2 ft (fig. 2–6). Explanatory time-series components that were used to filter out non-aquifer-test influences were barometric-pressure change and water levels from all four background wells. Some nonaquifer-test-related fluctuations in the value of synthetic water level minus measured water level remained after filtering; however, water levels from UFA well 35Q070 during the aquifer test were adequately filtered of barometric-pressure change and long-term water-level trend. Recovery data were dominated by noise that could not be adequately filtered out, and as a result, the recovery data were not used for aquifer-test analysis.

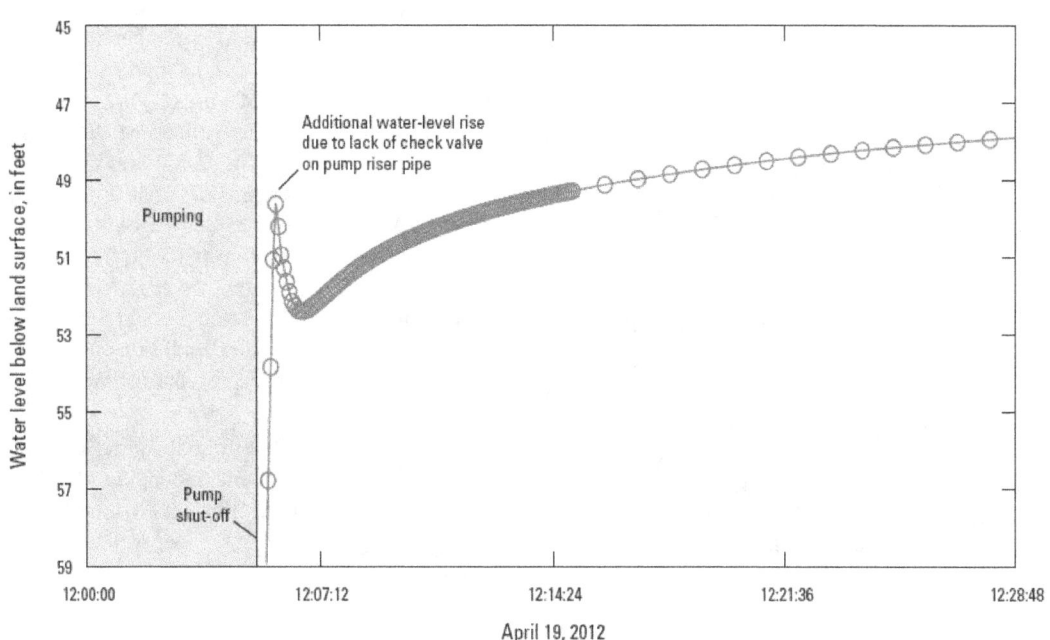

Figure 2–10.   Water level at the end of the 72-hour aquifer test in well 35Q069, Pooler, Georgia, April 19, 2012.

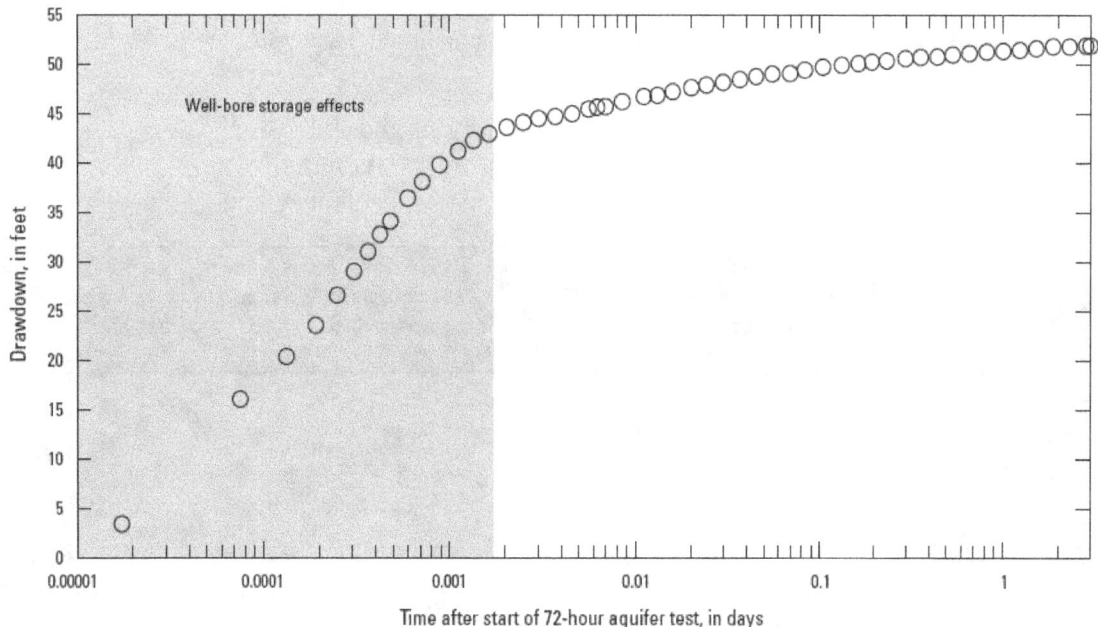

Figure 2–11.   Semi-log plot of drawdown in pumped well 35Q069 during the 72-hour aquifer test, Pooler, Georgia, April 16–19, 2012.

Model Simulation of the Lower Floridan Aquifer Test

Because of the nonlinear nature of drawdown on the semi-log plot, transmissivity of the UFA and LFA was estimated by using the numerical model MODFLOW-96 (McDonald and Harbaugh, 1988; Harbaugh and McDonald, 1996) with the calibration tool MODOPTIM (Halford, 2006b). Hydraulic properties were estimated by using the parameter-estimation program PEST (Doherty, 2005) minimizing the weighted sum-of-squares of differences between simulated and measured drawdown (hereinafter, these differences are referred to as "residuals"). Estimated drawdown, discussed in previous sections, is referred to as "measured drawdown."

The aquifer system was simulated with a two-dimensional, axisymmetric, radial, transient groundwater-flow model that incorporated LFA pumped well 35Q069 and UFA well 35Q070 (fig. 2–12; Harbaugh and others, 2000). Full description of the derivation of two-dimensional radial models using a single layer or multiple layers are provided in the following references: Rutledge (1991); Reily and Harbaugh (1993); and Langevin (2008). The method of computation of flow observations for parameter estimation with radial models is described in Clemo (2002). Horizontal hydraulic conductivity and specific storage were assumed to be isotropic and homogeneous within each hydrogeologic unit. Model simulation results were not sensitive to the vertical anisotropy ratio (vertical hydraulic conductivity divided by the horizontal hydraulic conductivity). A vertical anisotropy ratio of 0.1 was assumed for each hydrogeologic unit (table 2–4). A pumping rate of 783 gal/min was used in the aquifer-test simulations.

The model domain was discretized into 120 rows representing the different aquifer thicknesses and 60 columns representing the radial distance from LFA pumped well 35Q069 to the external boundary (fig. 2–12). The model radially extends 200,000.22 ft from the center of well 35Q069, and represents the subsurface depth interval between 43.38 (the estimated water-table depth) to 1,162.06 ft. Radial grid spacing (column width) consists of two increments of 0.333 ft to represent the borehole of LFA pumped well 35Q069. Column width beyond the borehole increases by a factor of 1.293, from 0.02 ft adjacent to the well to 45,282 ft at the edge of the model. Each row height represents a vertical thickness of 9.639 ft for the simulated aquifers and convening confining units.

Hydrogeologic units are represented in the model as five layers (fig. 2–12):

- Layer 1, represents the surficial aquifer system, Brunswick aquifer system, and Upper Floridan confining unit, undifferentiated;

- Layer 2, represents the UFA;

- Layer 3, represents the LFCU;

- Layer 4, represents the LFA; and

- Layer 5, represents the lower confining unit underlying the Floridan aquifer system.

Storage is assigned separately to parts of layer 1. The top model row represents water-table conditions and was assigned a specific yield of 0.1, within the range of specific yield for unconfined aquifers (Freeze and Cherry, 1979). All other model rows in layer 1 represent confined conditions and were assigned a specific storage of 2.55×10^{-6} ft^{-1}.

Table 2–4. Hydraulic parameters used to simulate measured drawdown and recovery and estimated values of transmissivity and storativity for the 72-hour aquifer test at pumped well 35Q069, Pooler, Georgia, April 16–19, 2012.

[ft/d, foot per day; ft^{-1}, per foot; ft^2/d, foot squared per day; NA, not applicable]

Hydrogeologic unit	Layer	Model rows	Horizontal hydraulic conductivity (ft/d)	Specific storage (ft^{-1})	Vertical anisotropy (dimensionless)	Thickness (feet)	Transmissivity (ft^2/d)	Storage coefficient (dimensionless)
Water-table row	1	1	[a]0.491857	[a]1.000×10^{-1}	[a]0.1	9.639	NA	NA
Overlying confining unit	1	29	[a]0.491857	[a]2.550×10^{-6}	[a]0.1	279.531	NA	NA
Upper Floridan aquifer	2	19	252.768000	3.206×10^{-6}	[a]0.1	183.141	46,292	5.872×10^{-4}
Lower Floridan confining unit	3	19	3.563630	2.484×10^{-6}	[a]0.1	183.141	653	4.549×10^{-4}
Lower Floridan aquifer	4	35	11.935000	[a]1.190×10^{-7}	[a]0.1	337.365	4,026	NA
Lower confining unit	5	17	[a]0.100000	[a]2.370×10^{-6}	[a]0.1	163.863	NA	NA
Floridan aquifer system	2–4	73	NA	NA	NA	704	50,971	NA

[a]Not estimated but assigned to the model.

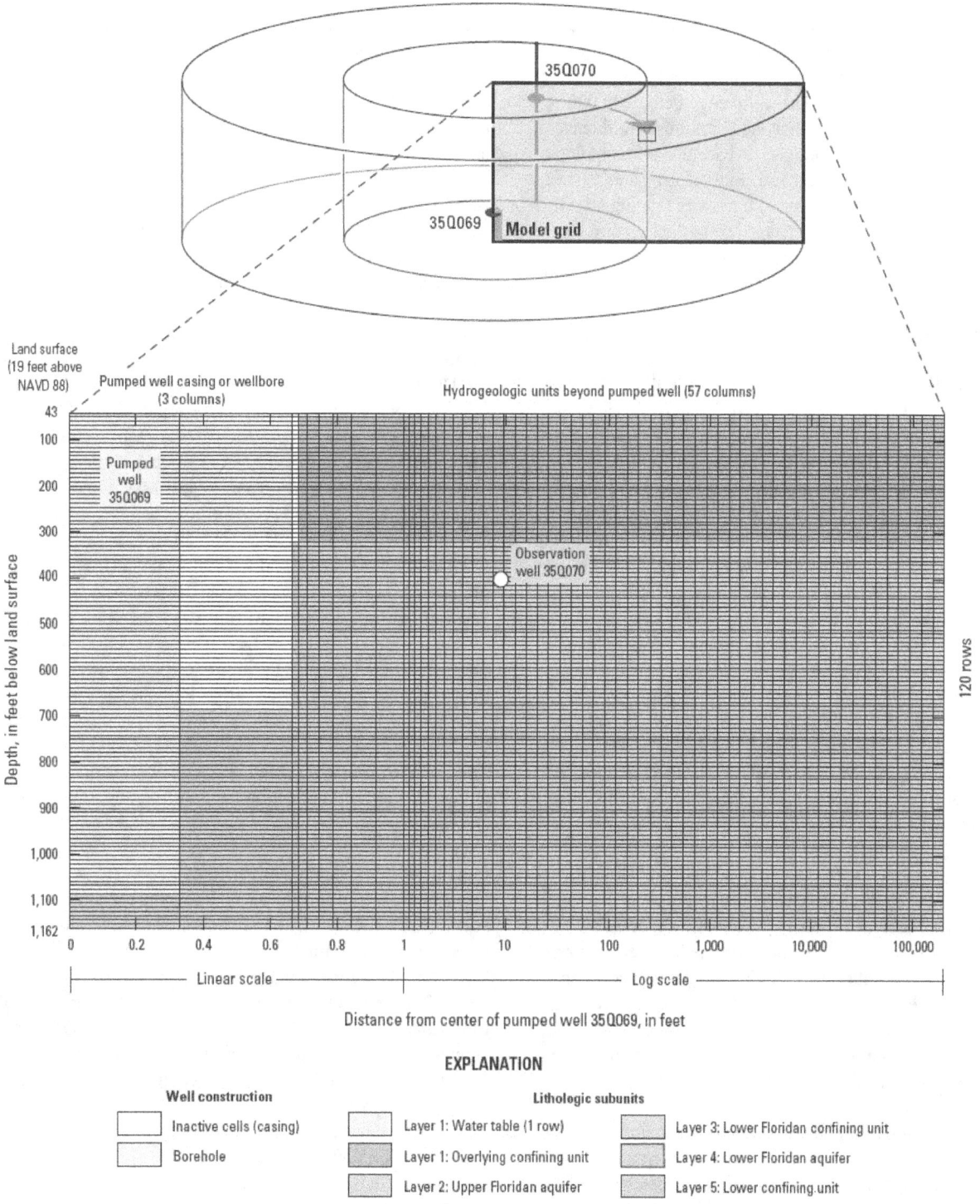

Figure 2–12.   Axisymmetric model for 72-hour aquifer test at pumped well 35Q069, Pooler, Georgia, April 16–19, 2012. Observation well 35Q070 and pumped well 35Q069 are included in the top diagram. Curved arrows in the top diagram represent transforming three-dimensional location of well opening to the two-dimensional model grid. The model is surrounded by no-flow boundaries.

The edges of the model are simulated as no-flow boundaries. This includes the proximal edge which is the center of the well, the lower boundary which is within the lower confining unit, the distal edge, and the upper boundary which represents the water table. The distal edge being 200,000 ft from the pumping well was beyond the radius of influence of the pumping well. The model is run with two stress periods that represent the aquifer test and recovery. The aquifer-test stress period is represented by 60 timesteps totaling slightly more than 3 days. Timesteps ranged from 0.92 second to 12 hours 52 seconds, with each succeeding timestep increasing from the previous timestep by a multiplier of 1.2. The recovery stress period is represented by 70 timesteps totaling slightly more than 5 days. Timesteps ranged from 0.3 second to 20 hours 2 minutes 24 seconds, with each succeeding time step increasing from the previous timestep by a multiplier of 1.2.

The model simulated the drawdown in response to the 72-hour aquifer test. No other influences were simulated so that initial heads and flow within the model were zero. During model simulation, water was injected at the same rate that water was withdrawn at the pumped well (783 gal/min). The water was injected into the model at the top cell representing the inner borehole (column and row 1) using the MODFLOW WEL package (McDonald and Harbaugh, 1988; Harbaugh and McDonald, 1996). High values for hydraulic conductivity and storage coefficient were assigned to the cells representing the borehole.

MODOPTIM computes the relative sensitivity of parameters and the "measure-of-redundancy" between parameters (Halford, 2006b). Parameter sensitivity indicates how adjustments to a parameter value will affect the objective function (sum-of-squares of the residuals or the goodness of fit between the simulated and measured data), and provides the basis for comparing simulated and measured drawdown. Parameters with very low sensitivity, relative to the most sensitive parameters (less than or equal to 0.01; Hill, 1998) are not estimated, but rather, are assigned a general value in the model.

The measure-of-redundancy is made between a pair of parameters. It describes how similar two parameters appear to a given objective function. A high measure-of-redundancy between parameters of the same hydraulic property usually means that one parameter should be estimated for multiple hydrogeologic units as a single merged hydrogeologic unit. A high measure-of-redundancy between parameters of different hydraulic properties (example, one parameter is horizontal hydraulic conductivity (K_h) and the other parameter is specific storage) means that one of the parameters should be eliminated from the estimation process. Values of measure-of-redundancy of greater than 0.99 indicate that the two parameters cannot be estimated independently (Poeter and Hill, 1997; Halford, 2006b). The relative sensitivities of parameters and values of measure-of-redundancy between parameter pairs in a MODOPTIM run were then used to determine which, if any, parameters should be eliminated from the estimation process and be assigned a general value in subsequent MODOPTIM runs.

The flowmeter-survey data indicated that no flow was coming from the lower confining unit in the bottom 80 ft of the open hole of well 35Q069. Assuming equal hydraulic head in all units and pumping during the flowmeter survey, no-flow contribution from a hydrogeologic unit equates to zero hydraulic conductivity for that unit. Both zero hydraulic conductivity and no flow are unlikely. The permeability of the lower confining unit is so much lower than that of the hydrogeologic units of the Floridan aquifer system, that a greater pumping rate would be required to observe water being produced from the lower confining unit. The K_h of the lower confining unit (layer 5) was not estimated in the final series of MODOPTIM runs but was set to a value of 0.1 ft/d.

The root-mean-square of the residuals divided by the maximum drawdown (RMS/MAX) was used to compare the success of runs in simulating well response to the aquifer test and recovery. The lower the RMS/MAX, the closer the simulated drawdown and recovery matched the measured drawdown and recovery. The residuals of the drawdown and recovery from LFA pumped well 35Q069 and the residuals of drawdown from UFA well 35Q070 were used to determine goodness of fit. MODOPTIM runs did not calibrate to the residuals of the recovery of UFA well 35Q070, because these data were poor due to failure to filter out nonaquifer-test influences as mentioned earlier in the report. Also, MODOPTIM runs did not calibrate to the residuals of about the first 10 minutes after the start and stop of pumping in LFA pumped well 35Q069 to avoid wellbore-storage effects during the start of the aquifer test and slug effects during the end of the aquifer test. Given that no estimated parameters have a relative sensitivity of less than 0.01 or a value of measure-of-redundancy with another estimated parameter of greater than 0.99, the values of parameters that make the best match (lowest RMS/MAX) of simulated response to measured response are the estimated values of those parameters.

The relative weights applied to residuals of UFA well 35Q070 compared to similar weights applied to residuals of LFA pumped well 35Q069 were adjusted in MODOPTIM simulations to best match the simulated response to measured response of both wells. The term "relative weight" means the weight on the residuals associated with UFA well 35Q070 relative to the weight on the residuals associated with LFA pumped well 35Q069. For example, a relative weight of 919 means that the weight applied to UFA well 35Q070 residuals was 919 times greater than the weight applied to LFA pumped well 35Q069 residuals. Initial MODOPTIM runs indicated that applying equal weights to the residuals of LFA pumped well 35Q070 and UFA well 35Q070 data led to a poor match of simulated drawdown to measured drawdown of UFA well 35Q070. The relative weight applied to UFA well 35Q070 residuals was increased in simulations that represented the 72-hour aquifer test in an attempt to calibrate MODOPTIM results for hydraulic flow properties of the UFA and LFA. Increasing the relative weight negatively affected calibrating to the LFA model parameters, and matching simulated drawdown and recovery to measured drawdown and recovery at LFA pumped well 35Q069. Increasing the relative weight applied to UFA well 35Q070 residuals affected net improvements in model accuracy in the UFA at the expense of actually decreasing the model's ability to accurately represent flow and aquifer-test water-level response in LFA pumped well 35Q069. An acceptable tradeoff was achieved where model improvement by increasing the relative weights associated with UFA well 35Q070 data outweighed the negative effects on model accuracy in LFA pumped well 35Q069 (and the model's ability to represent flow and water levels in the LFA).

Initial MODOPTIM runs provided preliminary estimates of parameters and determined which parameters should be estimated and which parameters should be assigned a set value. The K_h and specific storage of the confining units above and below the Floridan aquifer system (layers 1 and 5, respectively) were relatively insensitive, often had high measures of redundancy, and therefore, could not be estimated well with MODOPTIM runs. Values for these parameters changed little from initial values during the estimation process. In initial MODOPTIM runs, the value of specific storage of the LFA was estimated to be 1.19×10^{-7}, more than an order of magnitude lower than the value estimated for the other layers, and represents an unrealistic value for this aquifer parameter. The value for specific storage was set to its low preliminary value and was not estimated in final runs. Initial MODOPTIM runs also indicated that anisotropy within the Floridan aquifer system was not sensitive in the estimation process. Anisotropy was set to be 0.1 in all layers.

The final series of MODOPTIM runs was performed to estimate the K_h of the UFA, LFCU, and LFA (layers 2–4, respectively) and the specific storage of the UFA and LFCU (layers 2–3, respectively). Initial values of parameters used in

the final series of MODOPTIM runs were based on the results of initial MODOPTIM runs. Relative weights in favor of the UFA ranged from 1 to 3,676 (fig. 2–13). The base or starting value for relative weight was the ratio of the maximum drawdown of the LFA pumped well over the maximum drawdown of the UFA observation well (57.44). Other values of relative weight were 57.44 times 2^n, where n ranges from about −6 to 6. These multiples were used so that relative weights could plot on a log scale.

The final series of MODOPTIM runs indicated good matches of simulated response (drawdown and recovery) to measured response for relative weights from 1 to 919 (table 2–5). The only MODOPTIM runs that did not have good matches used relative weights of 1,838 and 3,676. The MODOPTIM run with a relative weight of 1 had a good match, because initial values of parameters based on the results from previous runs were already close to the values that provide the best match.

Parameter values were determined from the MODOPTIM run that used a relative weight of 105.35 (table 2–5). This MODOPTIM run has the lowest RMS/MAX without high values of the measure-of-redundancy. Despite the lowest RMS/MAX values for MODOPTIM runs with relative weights 114.88 to 919.04, a high value of the measure-of-redundancy between specific storage and K_h of the UFA and erratic fluctuations in parameter values precludes selecting results from this range of relative weights (fig. 2–13). The K_h values for the UFA, LFCU, and LFA were 253, 3.6, and 11.9 ft/d, respectively. These values round to 250, 4, and 12 ft/d, respectively. Specific storage of the UFA and LFCU were 3.2×10^{-6} and 2.5×10^{-6} ft^{-1}, respectively. These parameter values lead to a good fit between simulated and measure responses (fig. 2–14). Given the model thicknesses of the UFA, LFCU, and LFA (183, 183, and 337 ft, respectively), which are similar to actual thicknesses, the transmissivities of the UFA, LFCU, and LFA are estimated to be 46,000, 700, and 4,000 ft^2/d, respectively, and the values of storage coefficient of the UFA and LFCU are estimated to be 5.9×10^{-4} and 4.5×10^{-4}, respectively.

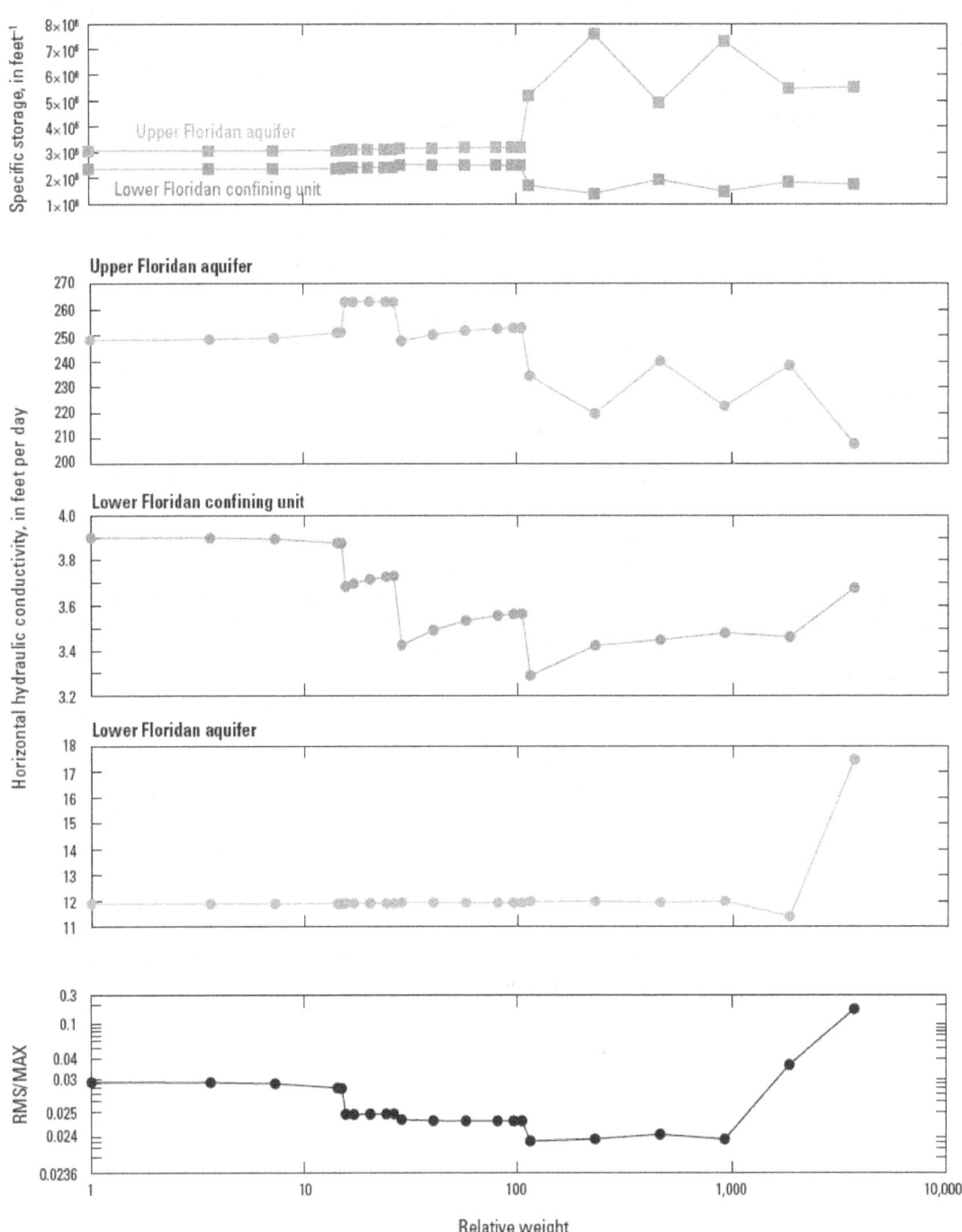

Figure 2–13.   Resulting parameter values from MODOPTIM runs and RMS/MAX as a function of the preferred weight of model calibration on the Upper Floridan aquifer observation well 35Q070 relative to the weight on the Lower Floridan aquifer pumped well 35Q069 (relative weight). Data are in table 2–4. RMS/MAX is the root mean squared of the differences between the simulated and measured drawdown values divided by the 72-hour (maximum) drawdown.

Table 2–5.   Resulting parameter values and values of RMS/MAX from MODOPTIM runs with different weights of calibration in favor of Upper Floridan well 35Q070, relative to Lower Floridan pumped well 35Q069, simulating response to 72-hour aquifer test, Pooler, Georgia, April 16–19, 2012.

[Relative weight, weight for model calibration of Upper Floridan well 35Q070, relative to the weight of Lower Floridan aquifer pumped well 35Q069; K_h, horizontal hydraulic conductivity; ft/d; foot per day; S_s, specific storage; ft^{-1}, per foot; RS, sensitivity of a parameter to model calibration relative to the most sensitive parameter. RMS/MAX is the root-mean-square of the difference in simulated and measured response, in feet, divided by the maximum drawdown in response to the aquifer test. Data used for the RMS/MAX were the drawdown, after wellbore-storage effects, and recovery, after back-syphon slug effects for well 35Q069 and drawdown for well 35Q070; UFA, Upper Floridan aquifer; LFCU, Lower Floridan confining unit; LFA, Lower Floridan aquifer]

Relative weight	Horizontal hydraulic conductivity $(K_h, ft/d)$			Specific storage (S_s, ft^{-1})		RMS/MAX	RS K_h UFA	Value of measure-of-redundancy of K_h UFA with other parameters			
	UFA	LFCU	LFA	UFA	LFCU			K_h LFCU	K_h LFA	S_s UFA	S_s LFCU
1.00	248.253	3.90132	11.9002	3.07×10^{-6}	0.000002	0.029259	0.010625	0.57	0.26	0.85	−0.36
3.59	248.430	3.89985	11.9004	3.07×10^{-6}	0.000002	0.029167	0.023713	0.06	0.12	0.89	0.12
7.18	248.979	3.89498	11.9012	3.07×10^{-6}	0.000002	0.028880	0.044367	−0.26	0.07	0.89	0.41
14.36	250.949	3.87748	11.9039	3.08×10^{-6}	0.000002	0.027927	0.086956	−0.57	0.06	0.90	0.60
15.00	251.169	3.87550	11.9042	3.08×10^{-6}	0.000002	0.027828	0.090737	−0.58	0.06	0.90	0.61
15.66	262.997	3.68467	11.9189	3.11×10^{-6}	0.000002	0.024878	0.091445	−0.59	0.06	0.90	0.61
17.08	262.875	3.69785	11.9177	3.11×10^{-6}	0.000002	0.024872	0.099827	−0.62	0.06	0.90	0.62
20.31	263.083	3.71622	11.9158	3.11×10^{-6}	0.000002	0.024893	0.118735	−0.67	0.06	0.90	0.65
24.15	262.959	3.72780	11.9146	3.11×10^{-6}	0.000002	0.024901	0.141242	−0.72	0.06	0.90	0.66
26.34	262.874	3.73137	11.9143	3.11×10^{-6}	0.000002	0.024902	0.154053	−0.74	0.07	0.90	0.67
28.72	247.781	3.42567	11.9472	3.15×10^{-6}	0.000003	0.024584	0.16483	−0.77	0.07	0.90	0.68
40.62	250.212	3.49150	11.9398	3.17×10^{-6}	0.000003	0.024531	0.23294	−0.82	0.09	0.90	0.70
57.44	251.757	3.53371	11.9360	3.18×10^{-6}	0.000002	0.024520	0.328515	−0.84	0.12	0.90	0.71
81.23	252.541	3.55579	11.9342	3.19×10^{-6}	0.000002	0.024518	0.457878	−0.86	0.16	0.90	0.71
96.60	252.744	3.56194	11.9346	3.20×10^{-6}	0.000002	0.024517	0.541934	−0.86	0.19	0.90	0.71
105.35	252.768	3.56363	11.9350	3.21×10^{-6}	0.000002	0.024517	0.589144	−0.86	0.20	0.90	0.71
114.88	234.224	3.28968	11.9940	5.21×10^{-6}	0.000002	0.023915	0.619642	−0.88	0.24	0.91	0.66
229.76	219.559	3.42362	11.9930	7.57×10^{-6}	0.000001	0.023956	1	−0.89	0.45	0.92	0.63
459.52	240.028	3.44858	11.9446	4.91×10^{-6}	0.000002	0.024061	1	−0.88	0.69	0.91	0.68
919.04	222.496	3.47916	11.9953	7.32×10^{-6}	0.000001	0.023952	1	−0.89	0.88	0.92	0.64
1,838.08	238.380	3.46175	11.3982	5.48×10^{-6}	0.000002	0.035769	1	−0.88	0.97	0.91	0.68
3,676.16	207.875	3.67805	17.4864	5.53×10^{-6}	0.000002	0.172299	1	−0.89	0.99	0.91	0.68

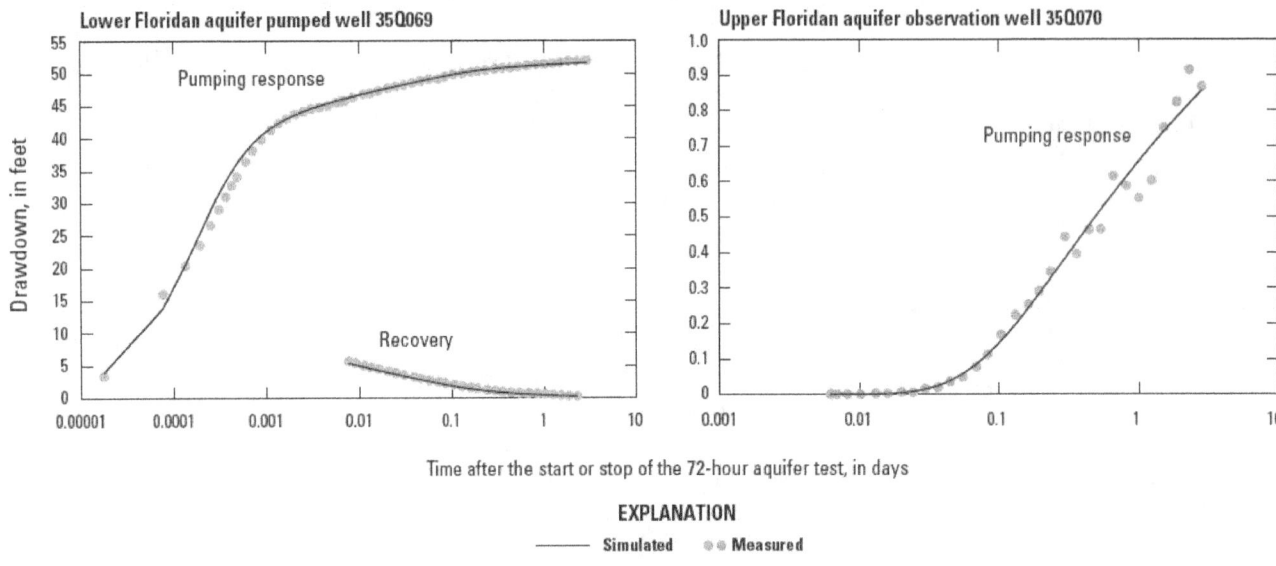

Figure 2–14. Simulated and measured water-level drawdown (pumping) and recovery at Lower Floridan aquifer pumped well 35Q069 and Upper Floridan aquifer observation well 35Q070 for the 72-hour aquifer test at pumped well 35Q069, Pooler, Georgia, April 16–19, 2012.